S0-ANQ-939

FUNDAMENTALS OF STEREO SERVICING

FUNDAMENTALS OF STEREO SERVICING

Joel Goldberg, Ph.D.

PRENTICE-HALL, INC., Englewood Cliffs, New Jersey 07632

Library of Congress Cataloging in Publication Data

Goldberg, Joel (date)
Fundamentals of stereo servicing.

Includes index.
1. Stereophonic sound systems—Repairing.
I. Title.
TK7881.8.G64 1983 621.389'334'0288 82-13340
ISBN 0-13-344549-6

Editorial/production supervision: Nancy Milnamow
Cover design: Mario Piazzo
Manufacturing buyer: Gordon Osbourne

Printed in the United States of America

10 9 8 7 6 5 4 3 2 1

ISBN 0-13-344549-6

Prentice-Hall International, Inc., *London*
Prentice-Hall of Australia Pty. Limited, *Sydney*
Editora Prentice-Hall do Brasil, Ltda., *Rio de Janeiro*
Prentice-Hall of Canada Inc., *Toronto*
Prentice-Hall of India Private Limited, *New Delhi*
Prentice-Hall of Japan, Inc., *Tokyo*
Prentice-Hall of Southeast Asia Pte. Ltd., *Singapore*
Whitehall Books Limited, *Wellington, New Zealand*

Contents

Preface

The solution to almost all electronic service problems lies in the complete understanding of how systems and circuits function. The technician of today and tomorrow needs to be able to analyze systems to localize an area of fault. Systems seem to be constantly changing. The introduction of the integrated circuit has led to the development of some very sophisticated circuits. Originally, the IC technology was limited to military and space developments. In the last several years this technology is found in almost all consumer electronic products. This new technology has forced the service technician to revise his or her thinking about repairs of all systems. One has to return to a solid understanding of how the system functions. A thorough knowledge of the blocks used in the system is required. In addition, one needs to know how signals are processed by the blocks as well as the required component circuit analysis.

The knowledge of the operation of systems is in addition to the knowledge of how and when to use test equipment. If one has a fear of using the oscilloscope, this fear must be overcome. Many of us are really afraid to use the scope because we never learned how to use it properly. It is one of the most versatile pieces of test equipment that is available today. The same IC technology that has improved consumer products is used to reduce the cost of the scope. Dual-trace scopes are appearing more and more on the electronic service bench. They are becoming a necessity for the diagnosis and adjustments of the latest groups of consumer electronic products.

The purpose of this book is to present material related to an understanding of consumer stereo electronic equipment. Section I explains the basics of how these systems operate. Section II covers procedures for analysis and repair of these systems. Much of the emphasis is on how the circuits and systems operate. This knowledge is fundamental to the diagnosis of the systems and their ultimate repair.

The author is aware of an apparent lack of ability on the part of students enrolled in electronic programs to relate their electronic education to the diagnosis and repair of many types of electronic products. This awareness led to the writing of a series of books on electronic servicing. This book is the third in a series on this subject. Each one is based on the concept that one must have an understanding of the basic circuit function before attempting to diagnose and repair any system.

The production of any technical book cannot be accomplished without the aid of others. These people deserve recognition for their support for this project. My family's support and encouragement is paramount. My thanks to my wife for her typing and initial editing of the manuscript. Many thanks to the training directors and service managers of the several manufacturers whose material appears in graphic form in the book. In addition, without the enthusiasm of my editor and the publisher this book would not have been written. Many thanks to all of these and the others who have contributed in some way to the completion of this manuscript.

Joel Goldberg

FUNDAMENTALS OF STEREO SERVICING

Section I

THEORY OF STEREO OPERATION

This portion of the book is devoted to a discussion of the basic electronic theories used in the operation of stereo devices. The material covers block diagrams as they relate to the function of the device. In addition, there is coverage of signals found in the stereo unit. Specific reference is made to AM, FM, stereo, and audio devices.

The latter portion of this section covers the selection of test equipment. Basic principles for troubleshooting are also discussed. The final portion presents material on how to use modern test equipment successfully. Included in this section is a discussion of how and why to use a dual-trace oscilloscope.

Chapter 1

Introduction to Stereo

The electronics industry has taken some major steps during the past several years. These include the development of the transistor, color television, high-quality tape recorders, and integrated circuits. These developments have led to the introduction of major consumer electronic units. High-fidelity sound is the result of efforts on the part of manufacturers and the demand by consumers for realistic reproduction of audio information. Technical improvements led to the introduction of two-channel audio systems. These systems are called *stereo.*

Stereo sound is a more natural sound, because it reproduces sounds as they would normally be heard. A block diagram for a basic stereo system is shown in Figure 1-1. The system actually consists of two identical amplifiers and speakers. A two-channel audio signal is developed for the input to the amplifiers. Each channel contains information that is normally heard by the listener's ears. Right-channel information is developed from a microphone placed so that its primary pickup range is the right side of the room. Left-channel information is developed from a microphone placed so that it picks up sounds from the left side of the room. These two signals are amplified by the system. Each is reproduced by the appropriate speaker. Speaker placement is shown in Figure 1-2. The system illustrated is called *two-channel stereo.*

The audio industry is always attempting to develop better systems for the reproduction of sound. One such system is called *four-channel stereo.* This system requires four speakers. It also needs either a four-channel amplifier or a two-channel amplifier and a decoder. The decoder is used to

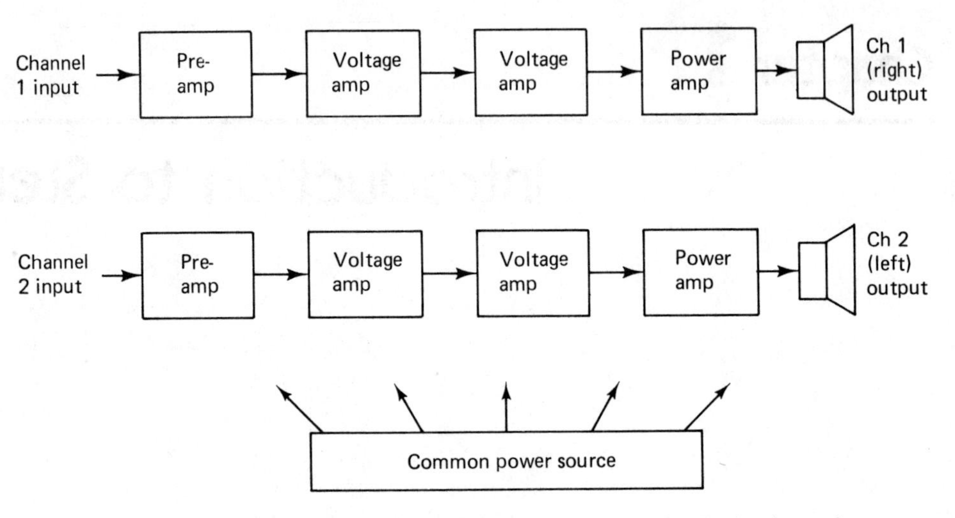

Figure 1-1 Block diagram for a typical stereo system. Both channels have identical characteristics. (From Joel Goldberg, *Radio, Television, and Sound System Repair: An Introduction*, © 1978, page 30. Reprinted by permission of Prentice-Hall, Inc.)

develop the audio for the third and fourth channels. These systems produce the slight echo effect one hears when listening to sounds in an auditorium. There is some "bounce" from the back of the room. These reflected sounds are also heard by the listener. Four-channel stereo was developed to reproduce this effect. Four-channel stereo speaker placement is shown in Figure 1-3. The system has not had the acceptance desired by its creators. Very few four-channel systems are available commercially at this time. Let us review the systems that are available for consumer use.

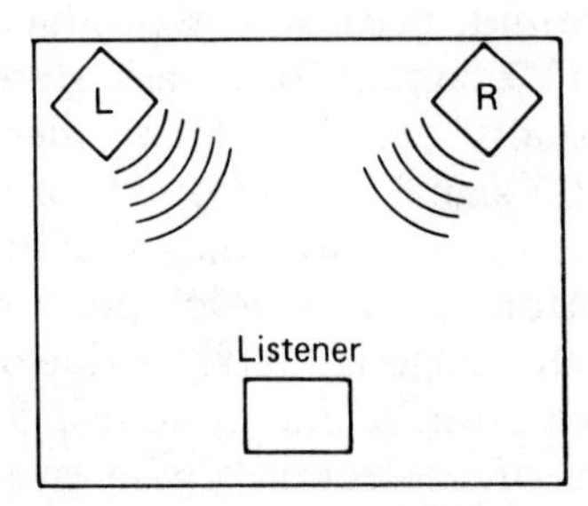

Figure 1-2 Speaker placement for a two-channel stereo system. (From Joel Goldberg, *Radio, Television, and Sound System Repair: An Introduction*, © 1978, page 31. Reprinted by permission of Prentice-Hall, Inc.)

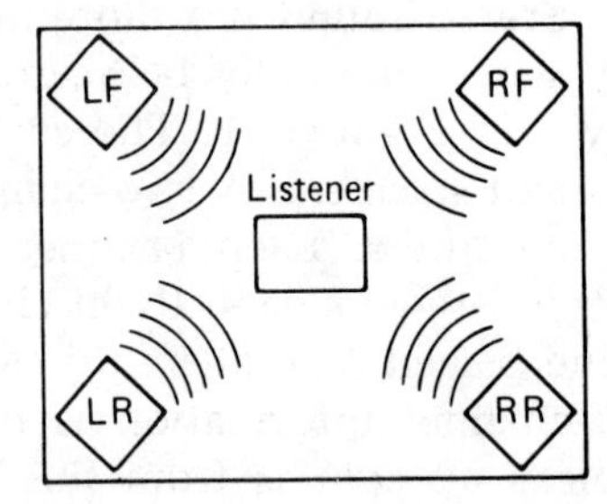

Figure 1-3 Speaker placement for a four-channel stereo system. (From Joel Goldberg, *Radio, Television, and Sound System Repair: An Introduction*, © 1978, page 31. Reprinted by permission of Prentice-Hall, Inc.)

THE STEREO SYSTEM

The development of the transistor and integrated circuit has led to the production of composite stereo systems. A system of this type is often packaged as a single unit. It contains several discrete subunits. In many cases the subunit selection is varied to meet the needs of the consumer. The building blocks of a complete stereo unit are shown in Figure 1-4. This unit contains several discrete blocks and has the flexibility to select a desired input to the system.

Each integrated stereo system is actually designed and produced by first selecting individual subunits. These subunits are then designed to fit into a housing. The end result is a package that contains all, or almost all, of the sections needed to reproduce sound. In some cases the input units are separate. The exact method of packaging the product depends on the manufacturer.

Each subunit has its own function, and each is discussed separately in this book. The actual wiring and physical connection information for a specific unit is given in the service literature for that unit.

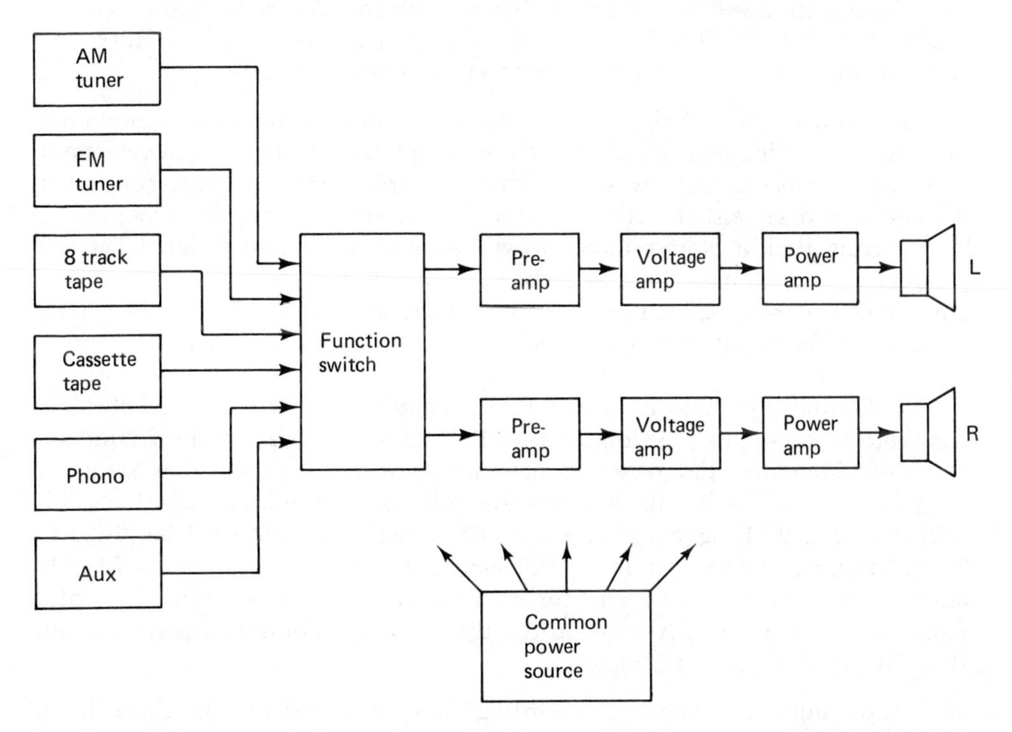

Figure 1-4 Block diagram for a complete stereo system, showing multiple input selection.

Inputs. Several typical input devices are shown in the illustration. These include AM radios, FM radio (stereo), two types of tape players, a phono input, and provision for auxiliary input devices. The desired input is selected by means of a function switch to the audio-amplifier system.

AM tuner. The AM radio input is capable of receiving, amplifying, and detecting an AM radio signal. Its reception frequency range is between 540 and 1600 kHz. The audio output from this system usually has an upper-frequency range of about 5 kHz. This is a monaural audio system. At the time of writing there is no acceptable AM stereo broadcast system available in the United States. A system has been proposed and accepted, but it was withdrawn by its proposers in 1980. Since this is a monaural system, the audio output from the AM unit is connected to both channels of the amplifier system.

By definition, a tuner is an electronic device that is capable of receiving and amplifying a radio-frequency signal. The tuner, as used in this definition, is a radio receiver that does not have any audio section. The output of the tuner is a demodulated audio signal. This tuner unit does not have the capability to amplify the demodulated audio signal.

Tuners are used in integrated stereo receivers. This reduces duplication of functional parts. It also permits switching of low-level signals. Both AM and FM tuners are used in the integrated stereo receiver.

FM tuner. The FM system is able to function in either a monaural or a stereo mode. This adaptability was mandated by the U.S. government when the stereo system was approved for broadcast use. Receivers must be able to process either a stereo or a nonstereo signal under this acceptance. This means that a stereo receiver will process a monaural signal and reproduce it as a monaural signal in both channels. It also means that a monaural receiver will reproduce a stereo signal in a nonstereo format at its output. Either receiver is also capable of reproducing the signal for which it was designed.

FM broadcast stations use the frequencies between 88 and 108 MHz for their carriers. The development of FM led to use of a broader frequency response. Monaural FM receivers have an audio output frequency range of 50 to 15 kHz. Additional information may be transmitted together with the monaural FM signal. This information requires additional bandwidth. Total available bandwidth for a full-service FM station extends to 75 kHz above and 75 kHz below the carrier frequency of the station. The total bandwidth is 150 kHz. A detailed discussion of the components making up this FM signal is given in Chapter 2.

Tape inputs. Magnetic recording tapes are available in three basic formats. The format described here refers to the method of recording in the tape and the means of storing the tape. The three storage formats discussed in this section are reel to reel, cassette, and eight track. All three systems are popular in today's sound market. A discussion as to which of the three systems is best is not relevant to the purposes of this book. Individual needs and personal biases determine this choice.

Selection of a specific tape transport or deck is left to the individual. The transport mechanism is strictly a mechanical device. Its purpose is to move the tape past a record or playback head in order to process information. Normally, the transport mechanism does not contain any electronics. All of the electronics required are found in the audio section of the system. Often, a transport mechanism does not have the capability of recording on the tape. It is used only for playback of prerecorded tapes.

The tape deck does contain electronics. It is used for both playback and recording. Its electronic system for playback is limited to a preamplifier. Additional audio amplification is provided by the audio amplifier section of the system. Provisions for selection of any of the tape systems depend on which units are built into the system. In some systems none are included. The system then has provisions for cable connections to the audio section.

Phono input. Another input is the phono connection to the system. Many different types of record players and changers are available on the market. Some of these are integrated into the system housing. Others are separate units. In either case selection of the audio signal from the phono is done by use of the function switch.

Auxiliary input. Many integrated systems have a provision for connection of an additional audio input. The selection of the specific input device is left to the user. This input, when selected, connects directly to the audio amplifiers. Inputs such as tape players are connected into the auxiliary input connections.

Function switch. The purpose of this switch is to connect electrically the desired input device to the audio amplifier system. Specifically, the output from one of the tuners or tape players or the phono is connected through the contacts on the switch to the audio-amplifier system. If the audio system is a stereo system, two sets of connecting wiring are required. One set connects to each amplifier channel. Under stereo operating conditions each input device also has two sets of connecting cables. In the AM mode, which is monaural, the inputs to both audio channels reproduce the same information when this is done.

Preamplifiers. The preamplifier is used when it is necessary to increase the level of the output signal from the tuner, phono, or tape player. The design of the audio-amplifier system components determines the amount of power that is available for operation of the speakers. This value can be exact only when a very specific amount of input signal is used. For example, one model of 40-W amplifier will provide the full output power when an input signal level of 500 mV is used. In this example, when the input signal is one-half of the 500-mV level, the output power of the amplifier is reduced to about 20 W.

Full output power for this model of amplifier requires that all input signals have the same strength. In many instances the assorted input devices have different signal levels. The ideal situation calls for the use of a preamplifier in order to have each input the same signal strength. In essence, this unit is an audio amplifier used to increase the input signal strength to a predetermined level.

The location of the preamplifier(s) often varies for different systems. Some manufacturers place preamplifiers before the function switch. More than one preamplifier may be required. Other manufacturers place the preamplifier after the function switch. A double section switch is used, as shown in Figure 1-5. Using the second arrangement only those inputs requiring preamplification are connected to this unit. Specific placement and application depend on the desires of the manufacturer. Often the service literature for a specific unit will include a functional block diagram for the unit. Review of this drawing will help to clarify the exact requirements for the unit.

Voltage amplifier. The electrical signal processed by the system requires much additional amplification. Signal levels from the input devices may be on the order of 500 mV and have a relatively low current of about 2 mA. This relates to 0.001 W of power (0.5 V × 0.002 A = 0.001 W). More amplification is required to develop the required output power for the system. A series of voltage amplifiers is used to increase the magnitude of this signal. Several stages of amplification are used because of the power design parameters of the specific devices used as amplifiers.

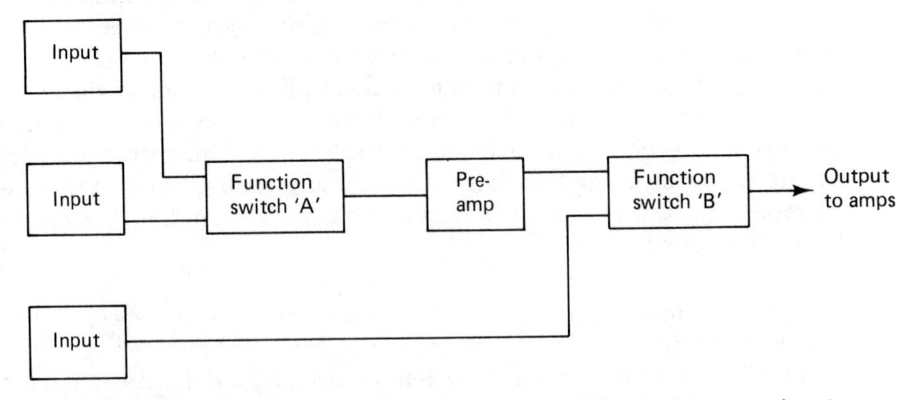

Figure 1-5 Placement of an audio preamplifier block to increase input signal levels.

Power output (amplifier). Output devices used to recreate sound waves require a relatively large amount of power. Most speakers use electromagnetic principles to create the desired sound waves. The electromagnets used in speakers are low-voltage, high-current devices. For example, the output-stage devices may develop a signal whose values are 40 V at 1 A of current. This will produce 40 W of output power (40 V × 1 A = 40 W). Most speaker systems in use today are not capable of handling the 40-V signal level. They also require more current than the 1-A value used here.

The power output system is also used as an impedance-matching unit. The speakers may require 40 W of power, but at a different ratio of voltage to current. The power output stage is designed to produce a value of 10 V at 4 A. This lower voltage and higher current still produces the required 40 W. The changed values are developed in the power output stage and match the requirements of the speaker system. The values of voltage and current used in these examples are not necessarily valid for a specific amplifier. Keep in mind that these figures are used only to illustrate how the system functions. Specific operating values depend on the design of the system.

Speaker system. The purpose of this system is to develop sound waves. There are several styles of speakers available. In addition, there are several types of speaker enclosures. Some speakers are designed to operate in only one portion of the audio-frequency spectrum. Often there are two or more speakers in one enclosure. Each speaker is used to reproduce a portion of the signal frequency. Electronic filters, called *crossover networks*, are used to direct the desired frequencies to the proper speaker. This subject is discussed in detail in Chapter 3.

Stereo system. The stereo system is a two-channel system. This requires two audio amplifier units at each stage. One amplifier system develops the left-channel sound and the second system develops the sound heard from the right channel. Both amplifier channels are identical in terms of their electronic circuits, gain, and components. This feature is helpful to the technician when service is required on the system. Very seldom do both channels fail. When one channel is working properly, it may be used to test the malfunctioning channel. Details of this procedure are discussed in Chapters 10 and 11.

Stereo decoder. A unit called an FM stereo decoder is used to process the FM stereo signal in the FM tuner. A block diagram showing the relationship of the decoder to the rest of the system is shown in Figure 1-6. When an FM stereo signal is processed by the tuner, components in the decoder separate the audio information into left- and right-channel information.

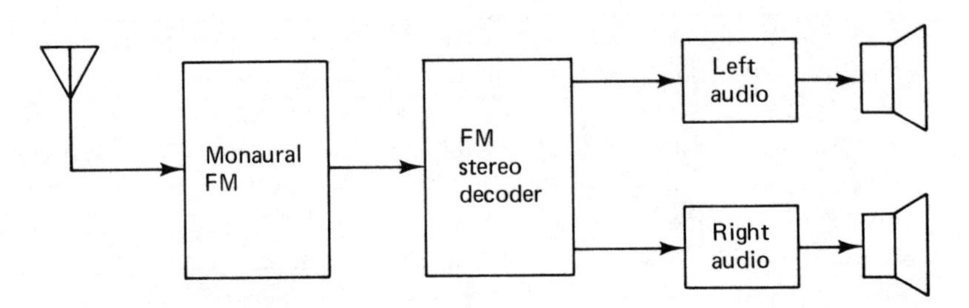

Figure 1-6 Placement of the stereo decoder block for an FM stereo receiver.

A special "turn-on" signal is transmitted along with the stereo portion of the information. Only when this signal is present will the decoder function. Without this signal, only monaural information is processed.

Noise reduction system. One problem that disturbs the serious listener relates to background noises, the hissing sound that is often present. This is particularly apparent with prerecorded tapes. Efforts have been made to reduce this disturbing noise without changing any of the quality of the sound. There are systems available that are added to the audio units which will effectively resolve this problem. A block diagram for one such system is shown in Figure 1-7. The specific electronic operation of these devices is discussed in Chapter 14.

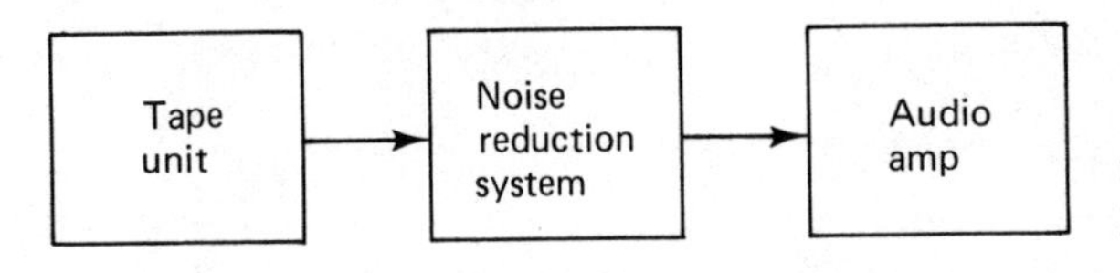

Figure 1-7 Placement of a signal noise reduction system for a tape playback system.

The preceding discussion provides a general background to stereo and to the systems used to reproduce audio information. Now, let us look at the present and the projected future of the audio-sound field. There are many new trends and developments occurring in the audio industry. All of these tend to produce better audio for the listener.

THE FUTURE IN AUDIO

The quality of audio available to the listener has improved greatly in the past few years. It will continue to do this over the next several years. Technological breakthroughs have produced much-better-sounding systems. Many radio stations are on the verge of broadcasting AM stereo signals. In addition, TV stations are preparing for FM stereo sound broadcasting. Digital encoding of the audio signal for recording purposes is presently being done.

Laser technology is being applied to reproduce information recorded on disks. All of these techniques, and more, will soon be in common use.

The development of new technologies has serious meaning for those people employed in the electronic service industry. First, it means that the technician of today must have knowledge of the processes used by the audio industry. Second, it means that if this technician wishes to be employable in the future, he or she must keep up with technical changes that are occurring. Third, the repair center owner must have new test and measuring equipment available for use with the newer circuits. The fourth major concern is that the technician be able to use this test equipment successfully—and not only use it successfully, but be able to analyze the results of the tests. In addition, the technician needs to know how to measure, remove, and replace components without damaging them in the process of repair.

All of these items are possible. Training and retraining will help service people to keep up with technical changes. A study of the several excellent articles that appear in technical journals will also be of great aid to the technician. Failure to stay abreast of the changing technology will lead to the inability to repair the products effectively.

Technological changes and innovative applications are occurring almost daily. It is almost impossible to keep up with every change that occurs. An excellent alternative is to develop good work habits and have a strong familiarization with how test equipment is used. These will go a long way toward helping one to be successful. The third required component for success is a thorough knowledge of electronic principles and their application. In the following chapters we will first review existing systems and then present material on how to service these systems. Also presented is introductory material related to some of the new technologies used in stereo systems.

QUESTIONS

1-1. Name three input systems used for stereo reproduction.

1-2. What is the frequency range of the AM radio broadcast band?

1-3. What is the range of audio frequencies used for AM broadcast?

1-4. What is the frequency range of the FM radio broadcast band?

1-5. What is the range of frequencies used for modulating the carrier of an FM station?

1-6. Name three types of tape systems.

1-7. What is the function of a preamplifier block?

1-8. What is the function of a power amplifier block?

1-9. What type of noises are processed in a noise reduction system?

1-10. What blocks are used in a stereo decoder?

Chapter 2

Stereo Receiver Analysis

It has been stated that life is really a series of judgments. Each of the judgments one makes is based on previous knowledge. These statements are certainly true when one attempts to evaluate the operation of a stereo system. A valid comprehensive evaluation cannot be done unless one has a reference upon which to make judgments.

The expert technician knows how each device being repaired is supposed to work. Use of the physical senses of sight, sound, smell, and feel help to evaluate the product. Judgments are based on recognizing component parts and comparing those in the unit to mental pictures of how these parts look when they are new. One also does a "nose" test to determine if components are too hot. Often a touch test will also help determine which components are running hot. Listening will help to determine where problems lie. All these evaluations are effective only when one knows how it all looks, feels, smells, and sounds when working correctly. The material in this chapter covers the operation of AM and FM tuners. This material should clarify any misconceptions about the theory and operation of these systems.

SIGNALS AND MODULATION

Tuners and radios process electronic signals. These signals consist of electromagnetic energy. They have certain specific characteristics. These characteristics relate to size, shape, duration, and position related to starting time.

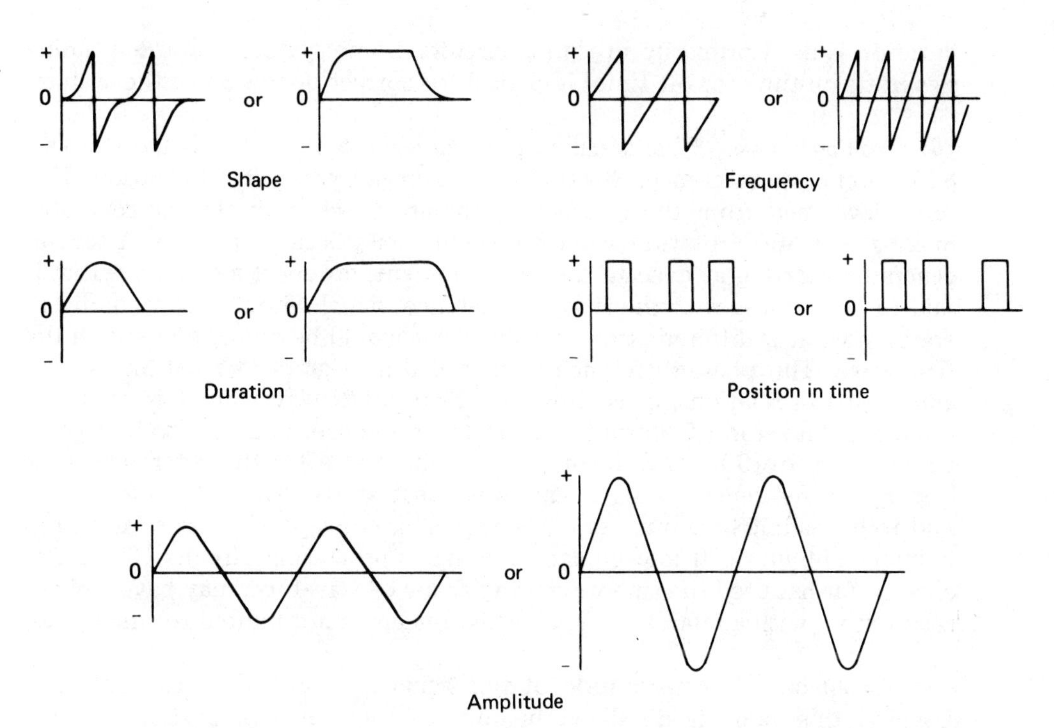

Figure 2-1 Characteristics of electronic signals include shape, time, and amplitude. (From Joel Goldberg, *Radio, Television, and Sound System Repair: An Introduction*, © 1978, page 87. Reprinted by permission of Prentice-Hall, Inc.)

Examples of signal characteristics are shown in Figure 2-1. All the signal forms used in electronic communication are modifications of three basic waveforms: sine, square, and sawtooth. Complex waveforms are developed from these basic forms when the factors shown in Figure 2-1 are employed.

Shape. All waveforms have their own recognized shape, or signature. This shape is observed on the screen of an oscilloscope. The form of the signal at the output of a system or circuit may not always appear to be the same as the wave at the input to the system. Electronic signals are processed through the system. Service literature for a specific device often provides waveform information for the service technician.

Frequency. This term is used to describe the quantity of complete repetitions of the electronic signal. A standard time reference of 1 second is used in the description. One complete repetition is called a *cycle*. The term *hertz* (Hz) is used to describe the number of cycles per second of the signal rate.

Duration. The length of time for a signal is called its duration. Signal duration is used primarily in timing circuits. Some systems require a timing signal. Its on time (or off time) is critical for correct operation of the system.

Position in time. This is called the *phase relation* of the signal. It uses a 360° circle as a reference. Starting time is considered as 0° of rotation. This term developed from the concept of the armature of an electric generator making one 360° rotation when it produces its basic sine wave. A second generator, starting at exactly the same moment, produces a sine wave that is said to be *in phase* with the first, or reference, signal. Should the second generator start at a different time, its sine wave would be *out of phase* with the first wave. The phase difference is described in degrees of rotation. An example of this relationship is shown in Figure 2-2. Waveform 1 is used as a reference. Waveform 2 starts 90° after the reference. It is said to be lagging waveform 1 by 90°. Waveform 3 is starting 45° after the reference. It is lagging the reference by 45°. Any wave that starts before the reference is said to be leading by a number of degrees. This concept of phase relationship is important in electronic control and signal-processing circuits. The reference so far has used the sine wave. Any shape of waveform may have a phase relationship with a reference signal. This concept is not limited to sine waves.

Amplitude. The magnitude of the signal is described as its amplitude. Usually, this value is discussed in units of the volt. Comparison is made from the highest point of the wave to the lowest point. This becomes a peak-to-peak (p-p) value. In some instances a zero-volt's reference point is included when describing the amplitude of the wave. This is illustrated in Figure 2-3. This signal has a +25-V portion as well as a −25-V portion. It

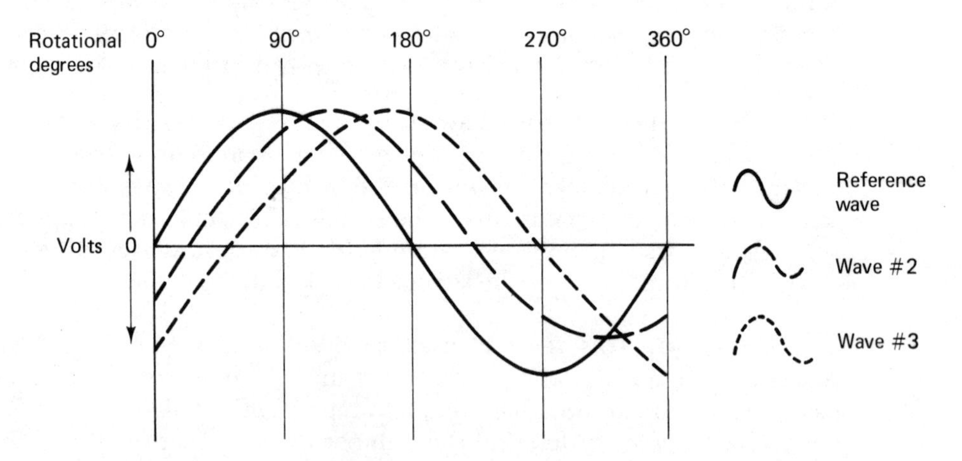

Figure 2-2 Phase relationship of signals refers to the starting time of the wave compared to a reference signal.

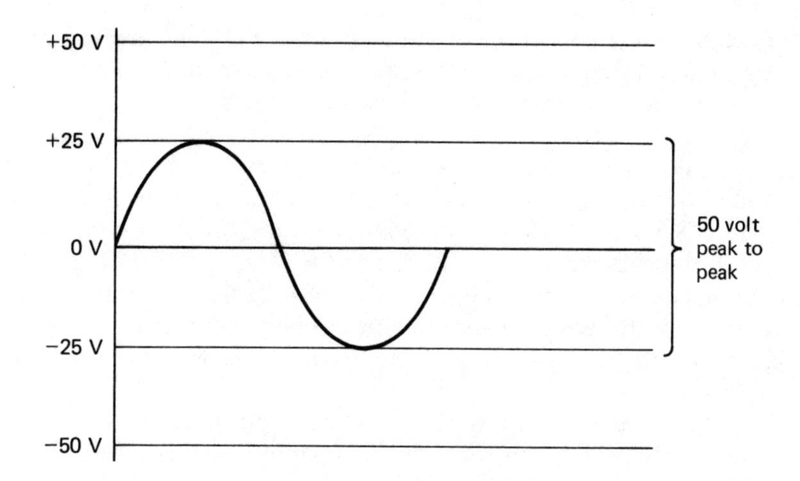

Figure 2-3 Electronic waves are measured in units of peak-to-peak voltage.

has a total amplitude of 50 V p-p. It can be described as having both a positive and a negative value, or as a peak-to-peak value.

Frequencies of waves. Electronic signals are also classified by their frequency range. Waves whose frequencies range between 10 Hz and 20 kHz are called *audio-frequency* (AF) *waves*. Others, with higher-frequency ratings, are called *radio-frequency* (RF) *waves*. The RF range has several subdivisions. These are classified into low, medium, high, very high, and ultra-high frequencies. The range of each division is shown in Figure 2-4. The exact value for each band edge is not critical and shows a slight overlap. The im-

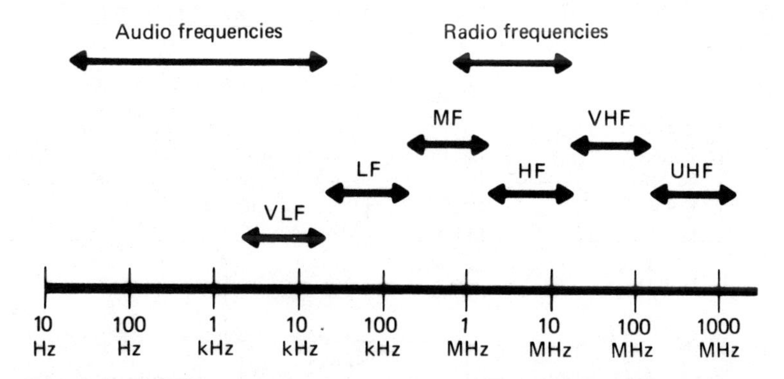

Figure 2-4 A general classification of radio waves is done according to their frequency. (From Joel Goldberg, *Radio, Television, and Sound System Repair: An Introduction*, © 1978, page 27. Reprinted by permission of Prentice-Hall, Inc.)

portant point is that all of those frequencies that are higher than the audio-frequency range are used to carry other electronic signals and also are used as electronic communication signals by themselves.

Carrier wave. Radio-frequency signals used for communications are called *carrier waves* or *continuous waves* (*CW*). These have a sine-wave shape. The CW signal frequency varies depending on the operational frequency of the broadcast station. In the United States, specific operational frequencies are assigned by the Federal Communications Commission (FCC). Each nation in the world has a specific agency to do this. Agreement among nations as to the use of specific frequencies is negotiated by international agreement.

Production of a CW signal is accomplished by use of a transmitter system. A block diagram for a simple system is shown in Figure 2-5. The carrier oscillator develops the basic signal. It then is amplified and frequency multiplied as required, using one or more RF amplifier stages. The transmitter system has a final RF amplifier stage. This stage is used to develop final RF power and to match the signal impedance to that of the antenna. In this manner maximum power is transferred from the final stage to the antenna radiating system. At the antenna the signal is converted into electromagnetic waveform energy and broadcast to the world.

The system illustrated does not contain any information at this time. Tuning a signal of this type on a radio will produce a quiet sound with little or no background noise. Communication can occur only when the signal is turned off and on as one would do to transmit using Morse code. If one wishes to use voice type of communication, an audio signal is added to the CW signal. This process is called *modulation*. There are several forms of modulation. These include two basic forms: amplitude modulation (AM) and

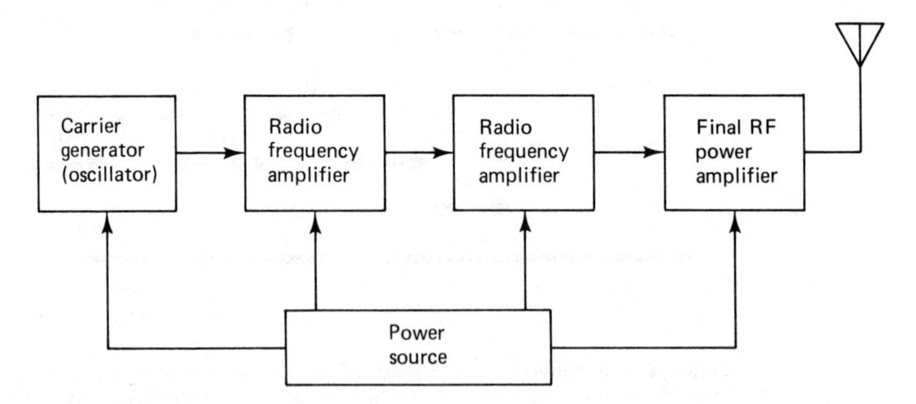

Figure 2-5 Block diagram for a carrier wave (CW) radio transmitter system.

frequency modulation (FM). Both of these are used in stereo broadcasting systems. It is important to understand how the signals are created because this knowledge is also used in the stereo receiver when the audio signal is reproduced.

Frequency modulation. A block diagram of an FM transmitting system is shown in Figure 2-6. This system uses the basic CW producing system to produce a *carrier*. The carrier waveform is that of a sine wave. The audio-frequency modulation will produce some changes in the shape of the waves. These changes are produced when the audio-frequency signals are electronically mixed with the carrier wave signals. The mixing of these two signals occurs in, or just after, the carrier oscillator block.

These changes, due to the modulating signal, cause a shift in the frequency of the carrier. An example of this action is shown in Figure 2-7. The upper waveform is that of the carrier. The waveform under it represents an audio signal. Its frequency is much lower than that of the carrier wave. The FM process produces the lower waveform. The modulating frequencies produce a shift in the carrier frequency. This shift has two dimensions. One of these is the amount of frequency shift that occurs. This deviation is dependent on the amplitude of the modulating signal. The other dimension is related to the number of times (frequency) of the deviation. The frequency of deviation is directly related to the frequency of the modulating signal. It is important to understand these factors because they relate directly to the systems used to demodulate the signal in the receiver. One last word on FM is that the ideal FM signal has a constant amplitude. This has provided for static and noise-free signal reception.

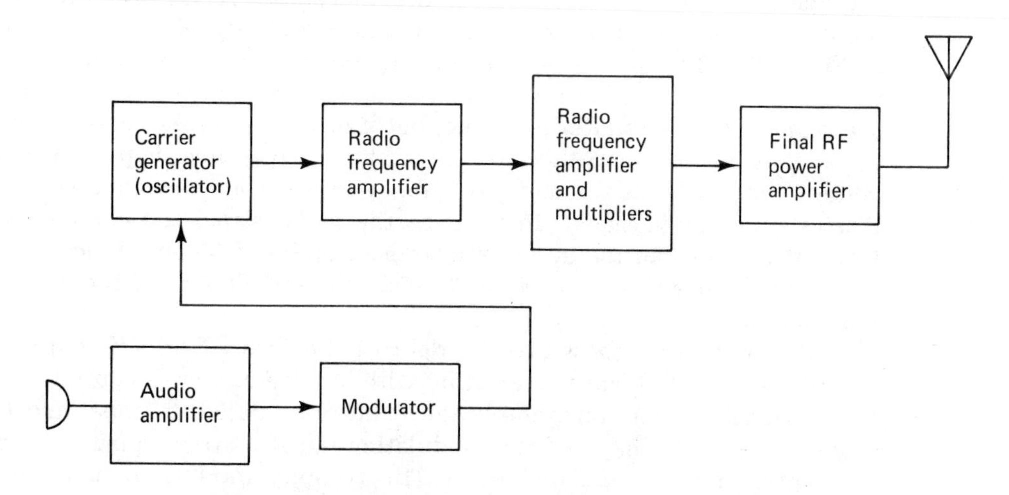

Figure 2-6 Block diagram for an FM radio transmitter system.

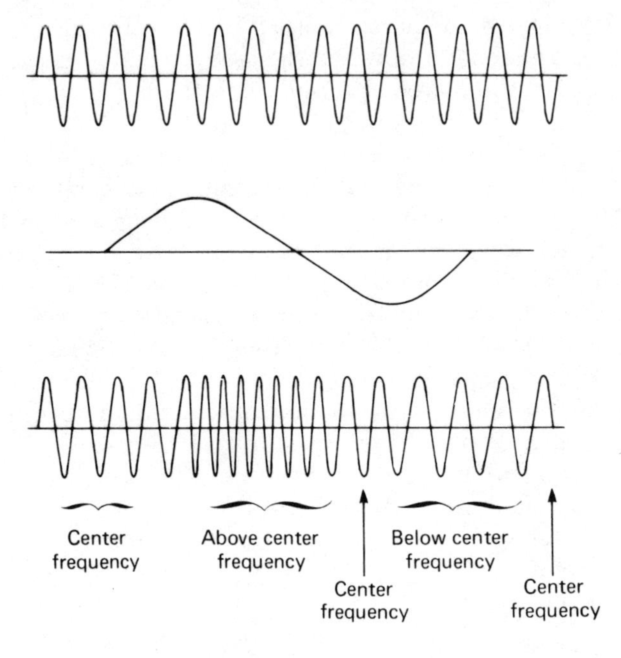

Figure 2-7 Frequency modulation changes the frequency of the station carrier wave. (From Joel Goldberg, *Radio, Television, and Sound System Repair: An Introduction*, © 1978, page 95. Reprinted by permission of Prentice-Hall, Inc.)

Amplitude modulation. A block diagram of an AM transmitter is shown in Figure 2-8. The major difference between FM and AM transmitters is the method of modulating the signal. Most commercial AM transmitters use a high-level modulation system. A separate power supply is used for the modulator and the final RF amplifier. The dc power for the final RF amplifier passes through the modulator on its way to the final stages. As it passes through the modualtor it takes on the shape of the modulating signal. This is illustrated in Figure 2-9. The pure dc is shown to the right of the power source block as a straight line, indicating a well-filtered dc source. The waveform shown to the right of the modulator is still dc, but it has taken on the shape of the AF signal that is applied to the modulator block. This, then, is producing a dc voltage that varies at the audio-frequency rate. A sine wave is used here to illustrate the audio signal. In practice the audio signal is a complex waveform. It contains all the audio frequencies required for communication. The sine wave is used here because it is simple and illustrates the factors of modulation.

The waveforms shown to the right of the final RF amplifier are the result of a modulated wave interacting with the RF carrier. The unmodulated RF carrier has a constant amplitude. Once modulated its upper and lower edges take on the shape of the modulating signal. Carrier signal is found in the center of the modulated wave. This complex modulated signal has the shape of the transmitted signal.

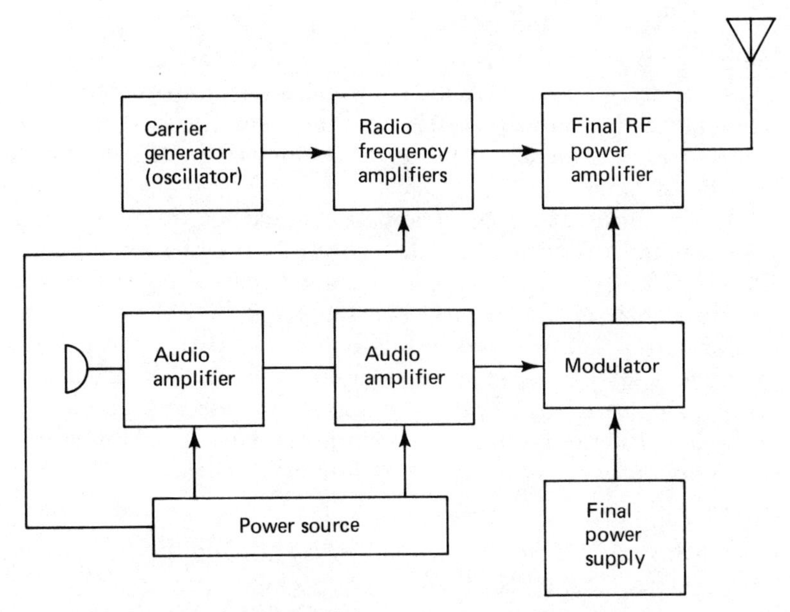

Figure 2-8 Block diagram for an AM radio transmitter. Note the placement of the modulator in this system compared to that of the FM transmitter.

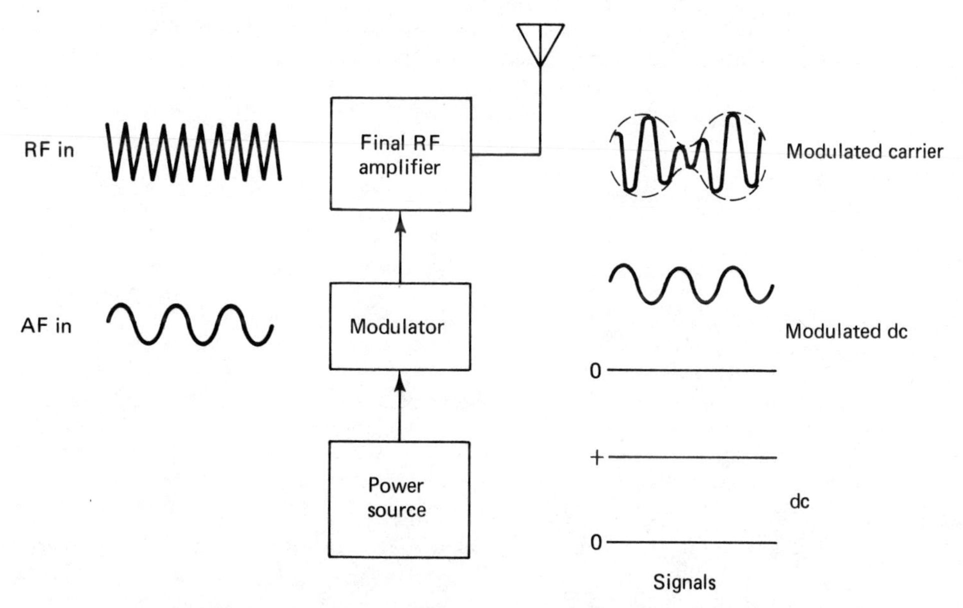

Figure 2-9 The shape of an amplitude-modulated carrier is determined by the modulating signal.

Bandwidth. One characteristic of any transmitted signal relates to the amount of space the signal occupies in the broadcast frequency spectrum. A pure carrier with zero modulation has a very narrow bandwidth. Modulating frequencies have both a positive and a negative factor. These factors add to the carrier frequency. This effect widens the space requirements of the broadcast signal.

The modulation signal adds to the carrier frequency. Since the modulation has both a positive and a negative factor, the modulated carrier actually develops both above and below the carrier. Thus a carrier modulated by a 5-kHz signal occupies a space totaling 10 kHz. The area occupied by this wider carrier is identified in Figure 2-10. In part (a) of this example a carrier frequency of 800 kHz is modulated by an audio signal whose frequencies vary from zero to 5 kHz. The space occupied by the modulated carrier is between 795 and 805 kHz, a total of 10 kHz. This is often shown as a spectrum curve, as seen in part (b) of the illustration.

Sidebands. The space occupied by the modulating portion of the signal is called its sidebands. Those frequencies that are higher than that of the carrier are called the *upper sideband* (USB). Frequencies below the carrier are called *lower sidebands* (LSB). Both types of sideband contain identical modulating information. There is, therefore, a duplication of signal being transmitted. Also, the modulated carrier occupies more space in the RF than is necessary for effective communication. The ideal situation is to develop a system that utilizes only a single sideband.

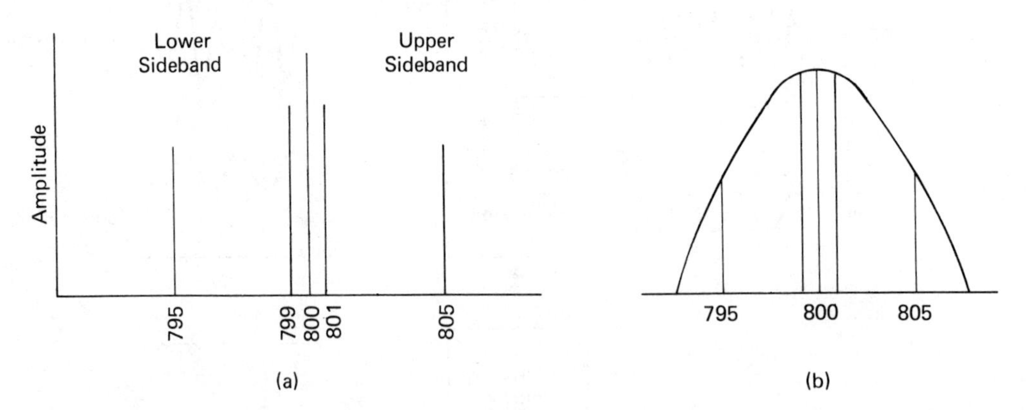

Figure 2-10 Sidebands are produced by the modulation of the carrier signal. Their width is equal to the upper frequency of the modulation. (From Joel Goldberg, *Radio, Television, and Sound System Repair: An Introduction*, © 1978, page 91. Reprinted by permission of Prentice-Hall, Inc.)

An analysis of the AM carrier signal shows that about 50% of the total energy used to broadcast the signal is used to develop the carrier. Each sideband requires 25% of the total power. In addition, each sideband contains both carrier and modulation information. It seems logical to utilize only a part of the total signal for effective communication. Several systems are in use that take advantage of these principles. One that relates to FM stereo broadcast is a double-sideband signal with a suppressed carrier. A spectrum waveform of this type of signal is shown in Figure 2-11. With this system the carrier signal is first created, then suppressed. The sidebands are amplified and broadcast. The FM stereo component signal does not require full broadcast power. The signal is added to the FM monaural signal using a system called *frequency multiplexing*.

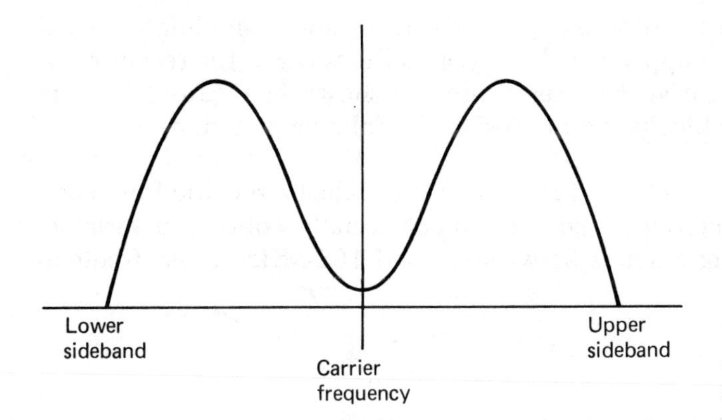

Figure 2-11 Spectrum analysis of a double-sideband suppressed carrier signal shows this form.

The FM multiplex system uses a carrier whose frequency is 38 kHz. This carrier is suppressed during its development in the transmitter. The result is a double-sideband suppressed carrier AM signal. This signal is the stereo component of the FM signal. Its sidebands have a bandwidth of 30 kHz. A spectrum analysis of the FM stereo spectrum is shown in Figure 2-12. The monaural portion of the signal, occupies the first 15 kHz of the spectrum. A reference signal, called a *pilot signal*, is transmitted at 19 kHz. The lower sideband of the stereo portion occupies the space between 23 and 38 kHz. The upper sideband's space is between 38 and 53 kHz. The 19-kHz signal is used as a reference in the receiver. Its frequency is one-half of the 38-kHz suppressed carrier. This carrier must be recreated in the receiver before demodulation of the stereo portion can occur.

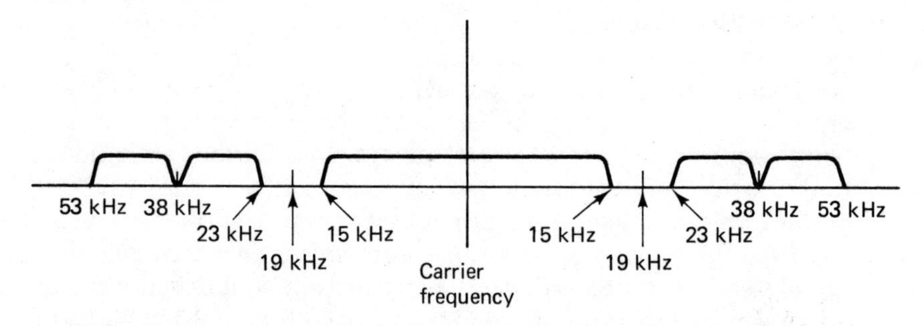

Figure 2-12 Spectrum analysis of an FM stereo signal, showing both upper and lower sidebands.

TUNERS

Once the systems used for developing the electromagnetic modulated signal are understood, one can apply signal analysis to the tuner in the receiver system. A block diagram for an FM stereo tuner is shown in Figure 2-13. The function of each of the blocks is described in the following section.

RF amplifier. The function of this block is to select a specific frequency and to amplify the received signal. This block usually contains a variable-tuned circuit. Its tuning range is between 88 and 108 MHz. These frequen-

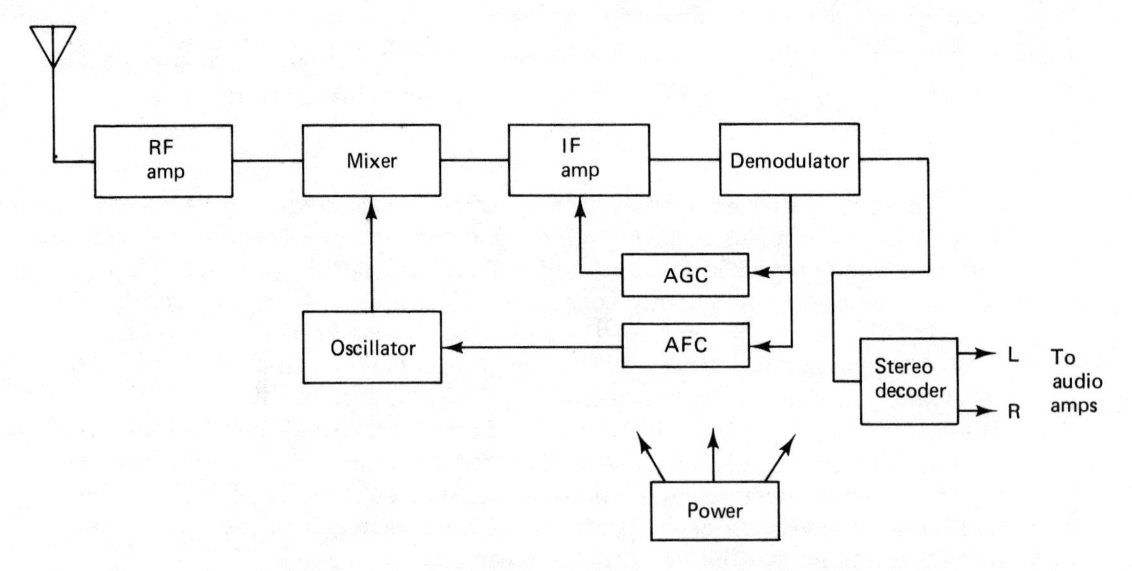

Figure 2-13 Block diagram for a heterodyne radio tuner system.

cies make up the commercial FM broadcast band in the United States. The RF amplifier has sufficient bandwidth so that it will amplify the complete FM stereo signal. This bandwidth allows both sidebands of the signal to be processed. About 10 to 15% of total signal amplification occurs in this block. The output of the RF amplifier block is fed to the mixer block.

Local oscillator. The local oscillator is used to create a CW signal. This block also contains a tuned circuit. Its tuning frequency range is between 98 and 118 MHz, about 10.7 MHz above the received frequencies. The local oscillator and RF amplifier tuned circuits are operated by the same tuning control. This permits both tuned circuits to "track" together. *Tracking* refers to the ability of one unit to follow another unit as its values change. A constant difference of 10.7 MHz is maintained between the RF amplifier and local oscillator blocks as they are tuned. The output of the local oscillator is fed to the mixer block.

Mixer. The function of this block is to accept two electronic signals and to mix them electronically. This will produce a third signal. In this type of system the third or output signal is one constant frequency. The principle of the mixer is to accept two signals that have different frequencies and to "mix" them electronically. The result is a great variety of signals. We are concerned with only one of them. It is the one signal whose frequency is the difference between the two input signals. Actually, four major signals can be identified at the output of the mixer. These are illustrated in Figure 2-14. The four output frequencies include both of the two input signals as well as both the sum and the difference of the input signal frequencies. Some receivers utilize the sum of the two frequencies. FM receivers use the difference of the two frequencies. This principle is called *frequency heterodyning*, or just heterodyning. For a number of years the advertising term "superheterodyne" has been used to describe receivers using the heterodyne principle.

The reason for using heterodyne principles is that they simplify and improve reception. Before this principle was introduced, each stage of a receiver had to be tuned by the listener. This required a person who was

Figure 2-14 Mixer action takes two input signals and develops four major signals from these.

willing to take the time to tune each stage carefully. Receivers were not as common as they are today. Signals from stations whose operating frequency was close to the station being received often interfered with the desired signal. The ability to select only the desired signal frequency was limited.

The heterodyne tuning system did away with these drawbacks. It permitted single-knob tuning and improved receiver selectivity.

Intermediate-frequency amplifier. This block, often called an IF amplifier, contains several tuned circuits. These tuned circuits are not adjustable by the listener. They can be adjusted by qualified service personnel as a part of an alignment procedure. The IF amplifier provides about 40 to 45% of the total signal amplification accomplished in a receiver. One-half of the amplification is found in the RF and IF amplifiers. The other half of the total amount of amplification is done in the audio-amplifier section of the receiver. We are looking only at the tuner section. The IF amplifier of the tuner, therefore, provides almost all of the signal amplification in the tuner section. The fact that this section is "fixed tuned" allows this amount of amplification to occur and eliminates the need for the listener to make any adjustments. The amount of amplification that occurs in this block is dependent on signal level. It is automatically controlled by another section of the tuner.

Control blocks. There are two sections of the tuner that are used as control blocks. One of these is called *automatic gain control* (AGC). Its purpose is to adjust the amount of amplification that occurs in the IF amplifier. This is a feedback type of system. A portion of the audio signal is taken from the demodulator, filtered, and returned to the IF amplifier. This voltage is used to bias the IF amplifier circuits. In effect it is used to control the gain of the various IF amplifier stages. Any changes in the level of the audio signal are quickly adjusted to a predetermined level. This helps maintain a constant-level signal at the demodulator.

The second control block is the *automatic frequency control* (AFC). This block also uses a feedback signal from the demodulator. It, however, is used to control the frequency of the local oscillator. All electronic components have a thermal characteristic. They tend to change value in relation to temperature. Frequency-sensitive circuits, such as oscillators, are designed with this in mind. The components used often are made in such a way that they maintain a specific value through a range of temperatures.

The AFC circuit is designed to correct for minor frequency changes that occur due to temperature variations. A control voltage is developed from an audio signal at the demodulator. This signal is filtered and returned to the oscillator circuit as a control voltage. Minor variations in oscillator frequency are corrected, keeping the oscillator on the selected frequency. Oscillator drift causes a detuning of the received signal.

Demodulator. This block is used to separate the audio information from the carrier signal. The output of the mixer is a carrier whose frequency is the IF frequency of 10.7 MHz. This carrier also contains the modulating signal developed at the transmitter. The sole purpose of the carrier is to bring the modulating information to the demodulator in an amplified form.

The FM demodulator reacts to changes in frequency of the carrier. It is designed so that it does not react to any amplitude changes. Many receiver demodulators have an amplitude-limiting circuit in order to delete any amplitude changes in the signal. The amount of frequency deviation and the rate of this deviation are used in the demodulator to recreate the modulating information. The amplitude-modulated FM stereo portion of the signal passes through the FM demodulator and is not demodulated at this time. Its amplitude is such that it is not affected by limiting action of the tuner.

Stereo decoder. This block is actually a complete system consisting of several subblocks. This breakdown is shown in Figure 2-15. When a stereo signal is being processed the decoder works in the following manner.

The monaural signal, identified as the L + R signal, passes through a low-pass filter. This filter is designed to allow only those frequencies below 15 kHz to pass through. Others are attenuated so that they are stopped before the output of the filter. The signal passes from the filter through a delay and then to both right and left matrix, or adder. We will put the signal on "hold" for a moment while we trace the stereo component paths.

There are two parts to the stereo signal. One of these is the 19-kHz pilot signal. The other is a double-sideband AM signal with a suppressed carrier of 38 kHz. The 19-kHz signal is amplified and sent to an oscillator. The oscillator's frequency is stabilized by the 19-kHz tone signal. From the oscillator the signal is frequency doubled to 38 kHz. It is then sent to a stereo demodulator block. It, too, will be placed on "hold" for a moment.

The second part of the stereo signal consists of an AM stereo component. This signal is identified as the L - R signal in the diagram. It is passed through a bandpass filter and amplified. This filter passes a band of signals whose frequencies fall between 23 and 53 kHz. These amplified sidebands are also sent to the stereo demodulator.

Both the 38-kHz carrier and the two sidebands are present at the stereo demodulator. Demodulation of the AM signal now occurs. The demodulator has two outputs. One of these is an in-phase signal. It is identified as the +(L - R) signal. The second signal is phase inverted by 180°. It is identified as the -(L - R) signal. Each of the demodulated signals is then sent to its appropriate adder, or matrix block.

The monaural L + R signal and the +(L - R) signals are electronically added at this time. The result is the formation of left-channel information.

Proof of this is the formula:

$$+(L + R) +(L - R) = \text{signal}$$

$$2L + 0R = \text{left signal}$$

At the same time the monaural +(L + R) is reacting with the phase-inverted -(L - R) signal. When the parentheses are removed, this becomes a -L + R signal. The combination of these two signals produces right-channel information:

$$+(L + R) + (-L + R) = \text{signal}$$

$$0L + 2R = \text{right signal}$$

Each of the signals is then sent to an audio-amplifier system.

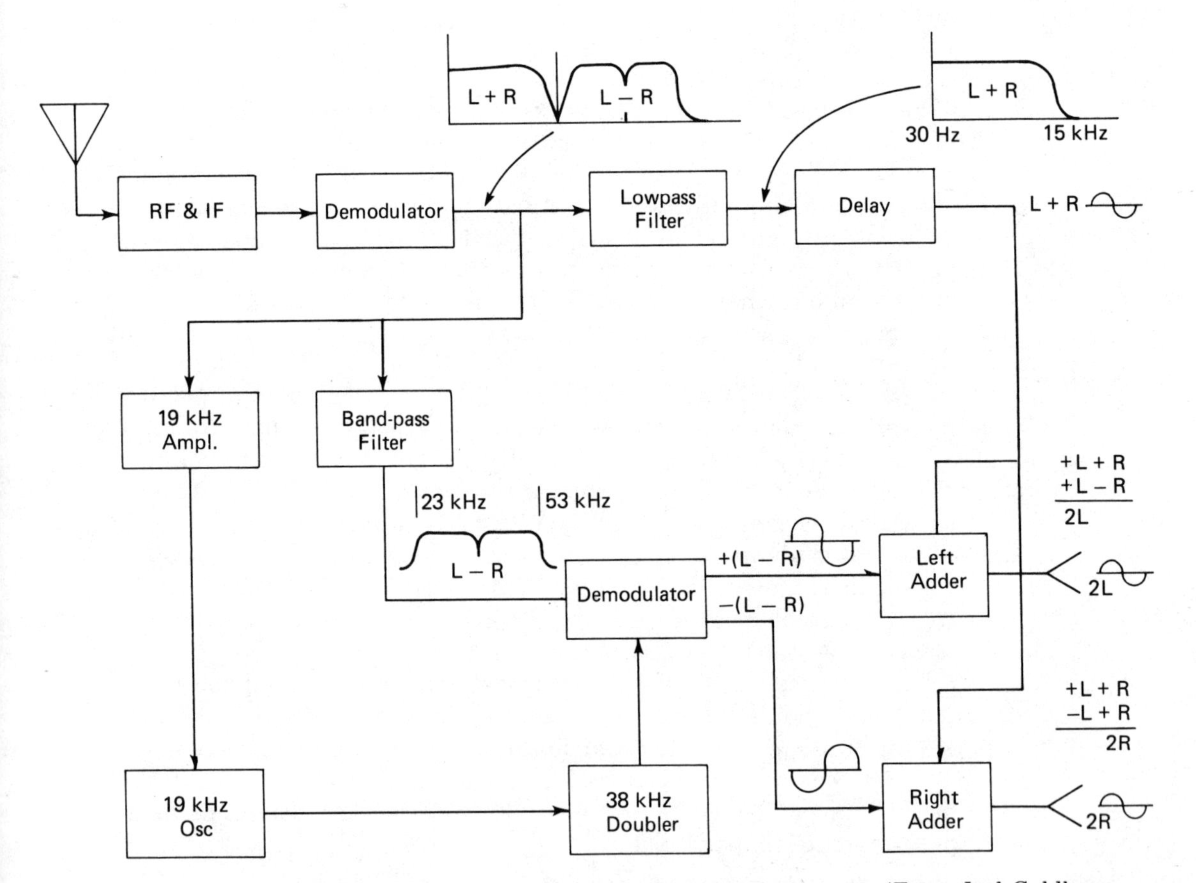

Figure 2-15 Block diagram for a stereo decoder system. (From Joel Goldberg, *Radio, Television, and Sound System Repair: An Introduction*, © 1978, page 130. Reprinted by permission of Prentice-Hall, Inc.)

Power source. Every electronic system requires a source of power. Specific values of voltage and current depend on the design of the system. Details related to the function and operation of power sources are covered in Chapter 9.

The other major section of the stereo system is the audio amplifier. This subject is so extensive that the entire next chapter is devoted to it.

QUESTIONS

2-1. What basic characteristics are used to describe electronic signals?
2-2. What is meant by the term "CW transmitter"?
2-3. What is the difference between an AM and an FM signal?
2-4. What changes occur to the carrier when it is amplitude modulated?
2-5. What changes occur to the carrier when it is frequency modulated?
2-6. What are sidebands?
2-7. What signals are found on a FM stereo broadcast station carrier?
2-8. Name the blocks commonly found in a FM stereo tuner.
2-9. What is the purpose of the 19-kHz tone found on the stereo signal?
2-10. What does the term "L - R" represent?

Chapter 3

Amplifiers

An amplifier by definition is a device that uses a small amount of power to control a larger amount of power. Electronic amplifiers meet this definition very well. There are several classifications for amplifiers. One of these identifies the frequencies of operation. These amplifiers are called RF, IF, VHF, and similar names. Another classification relates to the amount of power amplification that occurs. These are called low-power, medium-power, and high-power amplifiers. Still another classification is related to the method of biasing the amplifier. The biasing method determines the operating time characteristics. An additional classification describes the relative function of the amplifier. Very low level input amplifiers are called preamplifiers. Stages used between the input and the output blocks are called voltage amplifiers or just amplifiers. The final stage is called the power output or often just the output stage.

Each amplifier has its specific function in the order of things. The electronic technician needs to recognize an amplifier circuit. The technician also needs to be able to identify input and output connections for each type of amplifier circuit. In addition, one must be able to anticipate the approximate size and shape of input and output waveforms. Amplifier descriptions and functions are discussed in this chapter. Comprehension of this material will prepare one for the servicing of almost all audio-frequency amplifiers.

THE BASIC AMPLIFIER

One method of describing an amplifier is related to its input and output elements. Before this classification system is discussed, let us review the operation of a transistor amplifier. The circuit shown in Figure 3-1 is used to illustrate transistor operation. First, look at the dc operation of the device. Resistors R_{B1} and R_{B2} are used as a voltage-divider network. The voltage developed at this junction is called the *operating bias voltage* for the transistor. This circuit and the emitter–base circuit form the input circuit for the transistor. It uses about 5 to 10% of total circuit power.

The second circuit consists of the R_E, the emitter–collector terminals of the transistor, and R_L. This circuit has about 90 to 95% of the power consumption of the amplifier. If one is willing to accept the statement that the transistor emitter–collector junctions act as a variable resistance, it is fairly easy to understand how this device works. The output circuit has two fixed-value components. These are the emitter resistor R_E and the load resistor R_L. These two fixed resistors and the variable resistance developed between emitter and collector ($R_{E\text{-}C}$) form a series circuit. It is connected to the power source.

Ohm's law shows us the amount of current that flows in any circuit. If one knows the value of voltage and resistance in the circuit, the current can be determined by use of the formula $I = E/R$. If either resistance or voltage values change, the current will also change. The variable component $R_{E\text{-}C}$ thus controls current flow in the total circuit. Another application of Ohm's law shows the voltage drop that develops across a resistive component due to current flow through the device. Experience has shown that the size of the voltage drop is directly proportional to the size of the resistance. In other words, if $R_{E\text{-}C}$ changes in value, the voltage drops that develop across each of the three components in this circuit also change. This concept is illustrated in Figure 3-2. A circuit equivalent to the transistor output circuit is shown on the left side of this illustration. The graph on the right

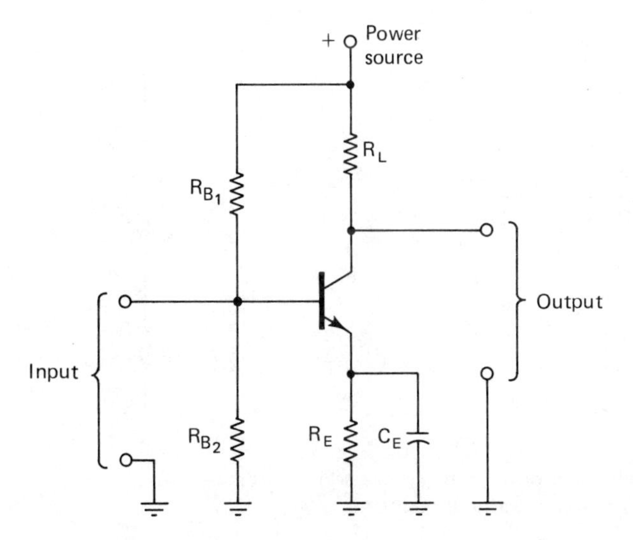

Figure 3-1 Schematic diagram for the transistor amplifier described in the text.

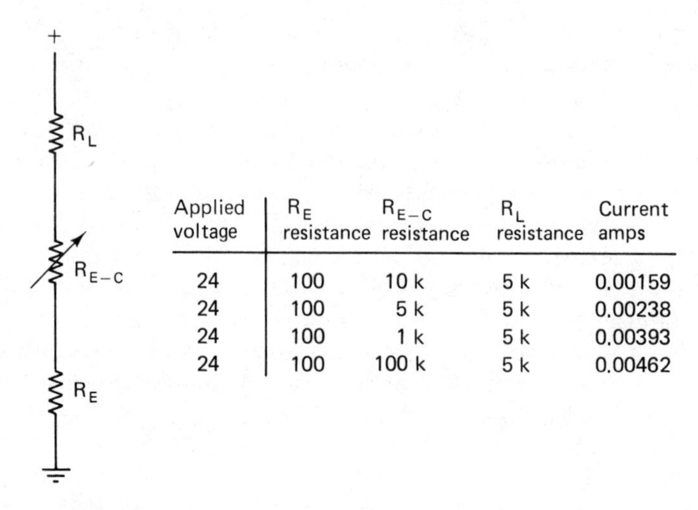

Applied voltage	R_E resistance	R_{E-C} resistance	R_L resistance	Current amps
24	100	10 k	5 k	0.00159
24	100	5 k	5 k	0.00238
24	100	1 k	5 k	0.00393
24	100	100 k	5 k	0.00462

Figure 3-2 A transistor output circuit acts as if there were two resistors connected in series.

shows current flow through the circuit for different values of transistor emitter–collector resistance. A voltage of 24 V is applied to the circuit to cause the current flow. The only resistance value that changes in this circuit is that of the emitter–collector junction of the transistor. As this value decreases there is a decrease in total circuit resistance. The result is an increase in total current flow.

Current-flow changes are only a part of the operating conditions of this transistor circuit. The change in current flow provides a change in the voltage drops that develop across each of the three resistive components in the circuit. This fact is shown in Figure 3-3. The voltage drops that

	Voltage drops		
Current	E_{RE}	$E_{E\text{-}C}$	E_{RL}
0.00159 A	0.159 V	15.9 V	7.95 V
0.00238 A	0.238 V	11.9 V	11.9 V
0.00393 A	0.390 V	3.93 V	19.65 V
0.00462 A	0.462 V	0.46 V	23.10 V

R_L

R_{E-C}

R_E

Figure 3-3 Voltage changes developed across series-wired resistors are shown as the middle resistance value is changed.

develop across each component are shown on the horizontal lines. The specific drops for each current flow are shown on the vertical column. The results of this show that the voltage drops decrease across the terminals of the transistor as current increases. Application of Kirchhoff's voltage law shows that the decrease in voltage across the transistor terminals is picked up as an increase in voltage drops across the other two resistors in the circuit. Also, the voltage drops increase as the transistor's resistance increases.

The result of this action is a dc voltage that develops between the collector of the transistor and circuit common. This dc voltage varies as the resistance of the transistor varies. This dc voltage is identified as the output voltage in this type of circuit. The dc voltage is changed by the action of the input circuit. In this amplifier configuration the input connections are between base and common. The output connections, as indicated, are between collector and circuit common. The circuit is called a common-emitter amplifier circuit because of the common signal terminals.

Small changes in signal conditions are able to control larger changes in the signal at the output of this type of circuit. An illustration of this is shown in Figure 3-4. The circle identified as "in" is a source for a signal. It is connected between the base and common. Its voltage adds to the fixed dc bias established at the base of the transistor by the transistor emitter–base junction and the 100-kΩ resistor. The result is a varying voltage at the base of the transistor. This voltage controls emitter–base current flow and the power at the input circuit.

The output conditions are shown to the left of the double line on the chart. Following any one line horizontally will provide operating conditions for the circuit. Maximum power output occurs at about the midpoint of the chart. The values shown are for dc voltages. The same principles apply

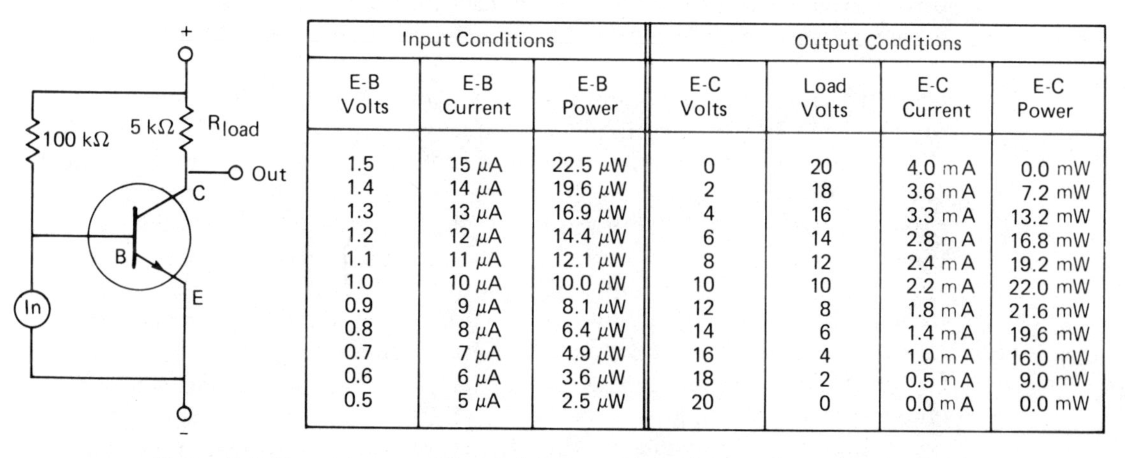

Input Conditions			Output Conditions			
E-B Volts	E-B Current	E-B Power	E-C Volts	Load Volts	E-C Current	E-C Power
1.5	15 μA	22.5 μW	0	20	4.0 mA	0.0 mW
1.4	14 μA	19.6 μW	2	18	3.6 mA	7.2 mW
1.3	13 μA	16.9 μW	4	16	3.3 mA	13.2 mW
1.2	12 μA	14.4 μW	6	14	2.8 mA	16.8 mW
1.1	11 μA	12.1 μW	8	12	2.4 mA	19.2 mW
1.0	10 μA	10.0 μW	10	10	2.2 mA	22.0 mW
0.9	9 μA	8.1 μW	12	8	1.8 mA	21.6 mW
0.8	8 μA	6.4 μW	14	6	1.4 mA	19.6 mW
0.7	7 μA	4.9 μW	16	4	1.0 mA	16.0 mW
0.6	6 μA	3.6 μW	18	2	0.5 mA	9.0 mW
0.5	5 μA	2.5 μW	20	0	0.0 mA	0.0 mW

Figure 3-4 Transistor amplifier and its operating values under various input conditions. (From Joel Goldberg, *Radio, Television, and Sound System Repair: An Introduction*, © 1978, page 223. Reprinted by permission of Prentice-Hall, Inc.)

for an ac input signal. The voltage at the base of the transistor varies as a result of the signal input and the operating bias. The output is a varying dc voltage that acts and looks similar to the input signal. There is a phase reversal in this circuit. Its causes are discussed in another section of this chapter.

In this amplifier circuit a small amount of power is used at the input. The power is in the microwatt range. The output circuit power in the milliwatt range. The statement that an amplifier is a device that uses a small amount of power to control a larger amount of power is valid for this circuit. The input circuit is the *control circuit.* The output circuit is the *controlled circuit.*

AMPLIFIER CLASSIFICATION

There are several ways in which to describe an amplifier. One of these, previously described, relates to the use of the amplifier in the circuit. These types are called audio, power, IF, and RF amplifiers. Another classification is used to describe input and output circuit connections. Still another way describes the relation of bias to the operating or duty cycle. The latter two methods are described in this section.

"Common" classification system. Transistors have three active elements. These elements are used in the description of the input and output connections. A chart showing some of these circuits is shown in Figure 3-5. The left-hand column shows a general type of circuit. The elements for this are identified as *origin*, *control*, and *sink*. These terms refer to the conducting flow through the device. The term *active device* refers to the fact that the device actively changes its internal resistance characteristic during its operation. The origin is the source of conduction. The sink is the element toward which the majority of conductors flow. The control is the element that has the ability to adjust or control the quantity of conductor movement.

Transistor polarity dictates the direction of electron flow through the device. This subject may be confusing to some. In the NPN type of transistor, electron flow is from emitter to base and collector. The PNP type of transistor has a current flow in the opposite direction. It goes from the collector and base to the emitter. In both types of transistors the conduction path for the majority of charges is from the emitter to the other two elements. To keep the discussion on a less confusing basis, this book looks at the movement of electrons rather than both electrons and the positive charges, or holes.

The upper drawing of the three uses the control element and the origin as the two input elements. A relatively small signal is applied between

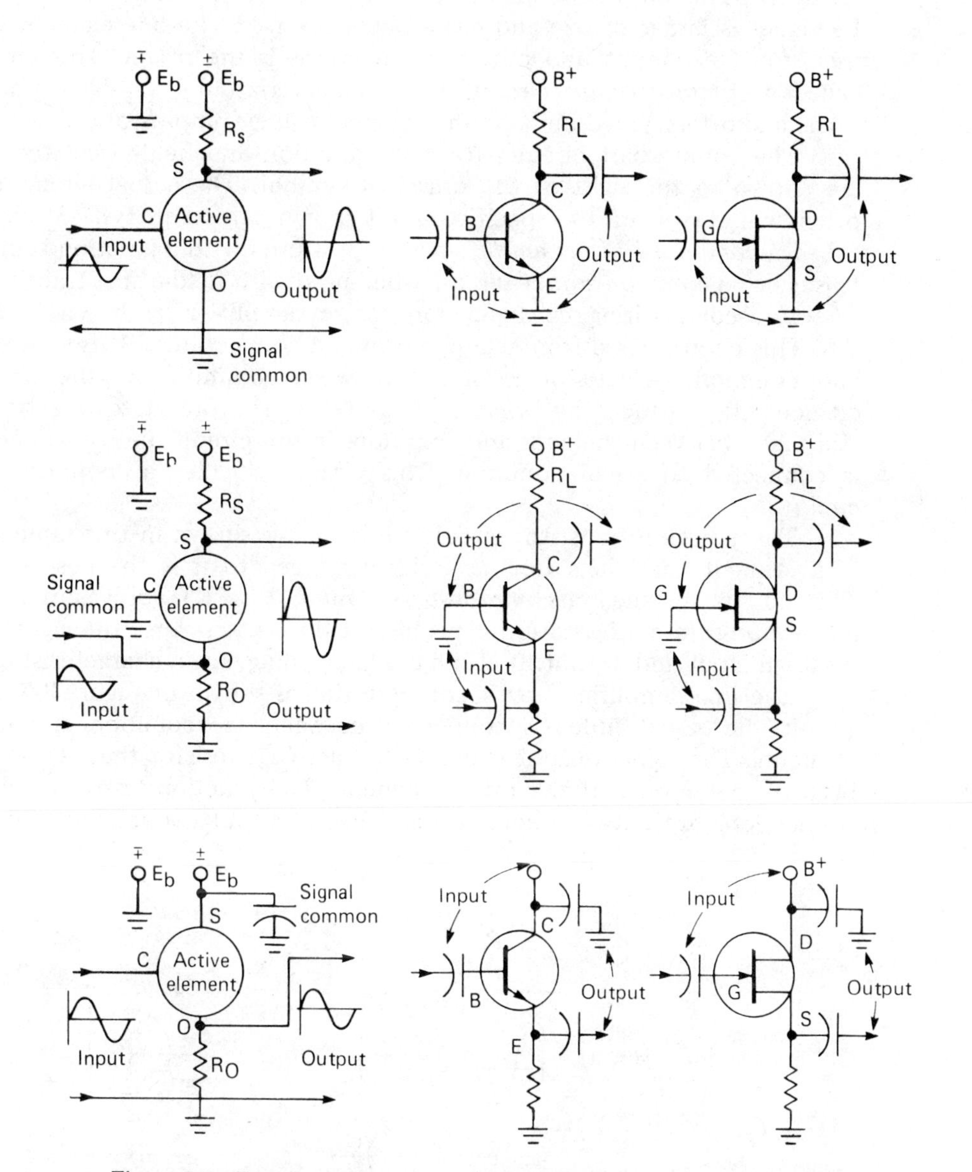

Figure 3-5 Classification system for amplifiers using the "signal common" identification method. (From Joel Goldberg, *Radio, Television, and Sound System Repair: An Introduction*, © 1978, page 227. Reprinted by permission of Prentice-Hall, Inc.)

these elements. It is a positive-going signal. It starts on the positive half of its cycle. The output elements are identified as the origin and the sink. Its signal is larger in size and has a phase reversal. The one element that is used for both input and output connections is the origin. This circuit is called a *common-origin circuit.* It is characterized by its signal phase reversal and offers a medium amount of signal voltage amplification.

The counterpart circuits for both junction and field-effect transistors are shown to the right of the universal symbol. The actual circuit design often will vary from the specific circuit shown here. This type of circuit is called a *common-emitter amplifier* when it is used with a junction transistor. It is called a *common-source circuit* when used with field-effect transistors.

A second wiring configuration for an amplifier is shown in Figure 3-6. This circuit has different input terminal connections. It uses the origin and common. A resistor is placed between common and the transistor element. It is called the *origin resistor* (R_o). The output connections are still made between the sink and common. In this circuit the control element is connected to signal common. This circuit is called a *common-control circuit.*

The junction transistor circuit equivalent is shown in this same figure. The element that does not have signal applied to it is the base element. This circuit is called a *common-base circuit.* Its FET equivalent is called a *common-gate* configuration. This circuit does not reverse the signal phase between input and output. It also has a large voltage-gain characteristic.

The final amplifier circuit configuration is shown in Figure 3-7. In this circuit the signal input is connected between the common and control elements. The signal output is connected across a resistor that is connected between common and the origin element. The junction transistor element connections are between common and base. Output connections are across

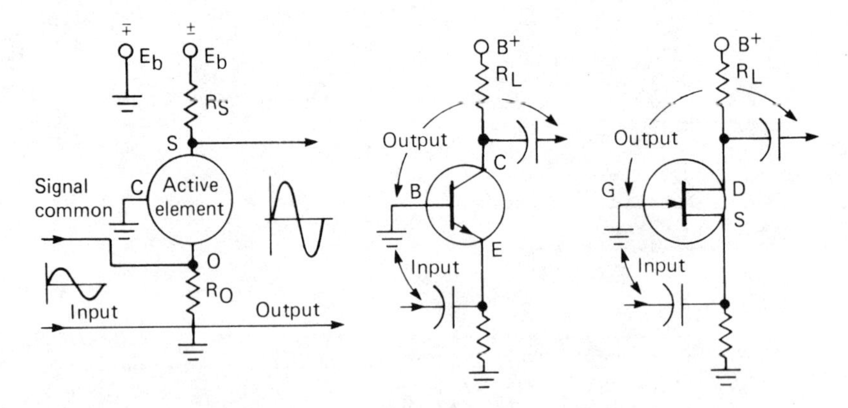

Figure 3-6 Common-control amplifier circuit. (From Joel Goldberg, *Radio, Television, and Sound System Repair: An Introduction,* © 1978, page 227. Reprinted by permission of Prentice-Hall, Inc.)

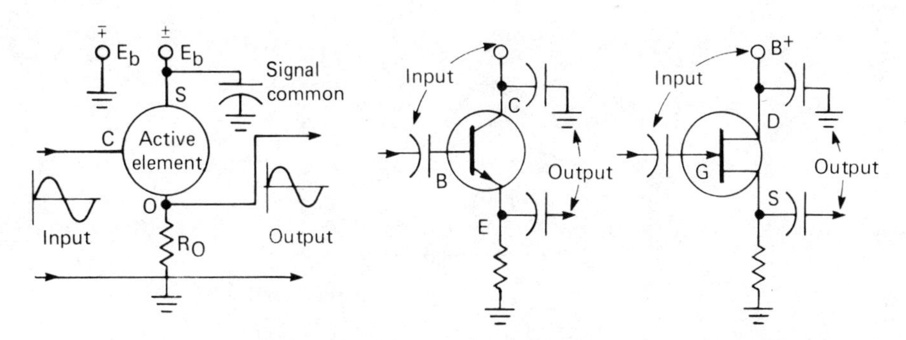

Figure 3-7 Common-collector or emitter-follower amplifier circuit. (From Joel Goldberg, *Radio, Television, and Sound System Repair: An Introduction*, © 1978, page 227. Reprinted by permission of Prentice-Hall, Inc.)

the emitter resistor. This circuit is called a *common-collector amplifier.* It is also called an *emitter follower* because the signal follows the emitter path. The FET circuit uses the same basic configuration. Signal input is between common and gate. Signal output is between common and the source resistor. The characteristics of this circuit configuration are no phase reversal of the signal and no voltage gain. Careful measurement of the output signal amplitude shows that it has about 98% of the input signal amplitude. Therefore, there is no voltage gain in this circuit. It does have a relatively high amount of current gain. This type of circuit is often used as a power output stage because of its power gain factor.

Signal and operating voltage. One item that often confuses many people is the fact that both operating voltages and signal voltages are present in all amplifier stages. These transistor devices cannot operate successfully without an operating voltage. This voltage establishes the static operating conditions for the amplifier. The introduction of a signal produces changes in these operating conditions. The net result of both dc operating voltages and the signal causes the resistance of the active device to change. The resulting resistance of the active device to change. The resulting resistance changes produce a variation in current flow. This, in turn, allows a varying dc voltage to develop at the output terminals of the device. The changing dc operating voltage takes on the shape of the signal voltage that produced it. This is illustrated in Figure 3-8. This uses a common-emitter amplifier circuit as the reference. The chart provides numerical data used to develop the graphs. The input signal voltage swing, or change, is 0.4 V. It starts at a bias level of 1.0 V, rises to a high value of 1.2 V, and then drops to a low value of 0.8 V. This is a signal input of 0.4 V peak to peak.

The input signal voltage controls the voltage between emitter and collector. With no signal applied, this voltage is 10V. When the input signal

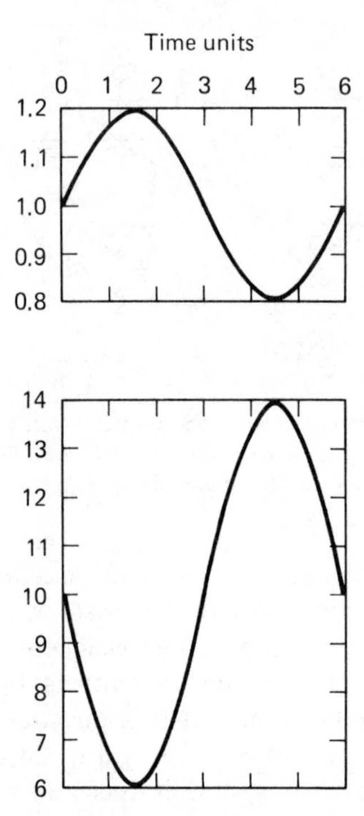

Figure 3-8 The dc output voltage at the collector of the amplifier assumes the form of the input signal.

rises, it causes an increase in conduction in the transistor. This also causes a decrease in the interval resistance and also a decrease in voltage developed across the output terminals of the transistor. The result is shown on the graph. Use the time units as a reference. The dc voltage measured at the output terminals of the transistor changes, as does the input voltage. This change is 180° out of phase with the input signal. It is also larger in value than the input signal. This output voltage swing of 8 V is called the *output signal.* In reality it is a varying dc voltage that is produced by changes at the input terminals of the transistor. These changes occur with any shape of input signal. This is the process called *amplification.*

A second amplifier circuit is shown in Figure 3-9. This is the common-collector, or emitter-follower circuit. The schematic drawing for the circuit is shown on the left side of the illustration. The upper graph shows the signal voltage. The lower graph shows the dc voltage that develops at the emitter of the transistor. The collector is connected directly to the source supply. A decoupling capacitor is used to remove any variations due to signal at this terminal.

The output circuit for this amplifier consists of two components. These are the emitter resistor and the emitter–collector terminals of the transistor. This is a basic two-component series circuit. One of these components, the transistor, is variable. The other, the emitter resistor, is fixed. In this series circuit, as in any other series circuit, the voltage drops depend on the size of each resistance. Changing the transistor resistance changes the voltage drops across its terminals. As these drops are reduced the voltage rises across the emitter resistor.

The output connections for this circuit are across the emitter resistor. Signal increases at the input, or base circuit, reduce the voltage drop across the terminals of the transistor. This produces an increase in the dc voltage between the emitter and circuit common. The input and output signals are always in phase in this circuit because of this action. The size of the emitter resistor determines its actual voltage drop. The changes that occur are unable to equal the amplitude of the input signal. They do, however, come very close to this value. Signal voltages at the output of this circuit will be about 95% of the value of the input signal voltage value.

The emitter-follower circuit does have a current gain factor. Consider the current factors in any transistor amplifier circuit. About 5% of the total current flow is in the base–emitter circuit. All the current occurs in the emitter. A small voltage drop across the emitter resistor will control a relatively large current flow through this resistor. A typical small-signal common-emitter amplifier will have about 10 μW of power at the input circuit. This is based on a voltage of 1.0 and a current of 10 μA (1.0 V × 0.00001 A = 0.00001 W). The output circuit will have 10 V and 2.2 mA. This is equal to 22 mW of power.

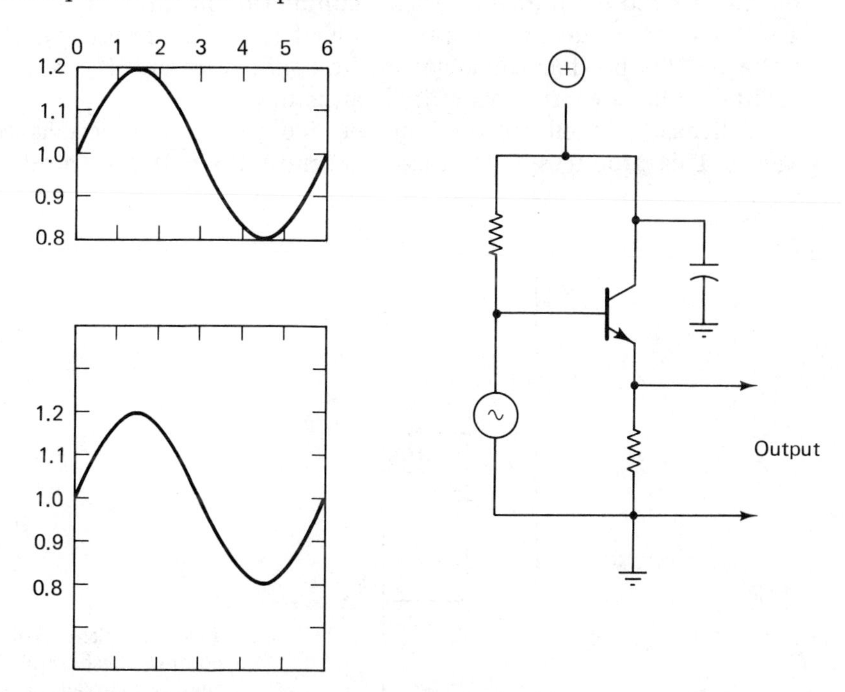

Figure 3-9 The dc voltage at the emitter of the transistor amplifier is in phase with the input signal.

Emitter-follower signal amplifiers are usually used in audio circuits as power amplifiers. Signal conditions at the input elements will be close to those described for the common-emitter amplifier. The output circuit usually has a much larger current flow. In some circuits the current will exceed 10 A. If the model circuit being examined had an output voltage of 0.95 V at 8 A, the output power is 7.6 W. This is equal to a power gain of 760,000. Gain is figured by using the formula

$$\text{Gain} = \frac{\text{output}}{\text{input}}$$

The current gain for this circuit is almost the same value. The figures used here are selected to show the factors relating to the gain and signal phase relation of the emitter-follower amplifier.

The final amplifier circuit is shown in Figure 3-10. This is the common-base configuration. Signal injection is between emitter and circuit common. Signal output is between collector and circuit common. The base is held, or clamped, at a fixed operating voltage by the decoupling capacitor in its circuit. All silicon junction transistors develop a base–emitter voltage of about 0.7 V when they conduct. This is a constant value. When the emitter voltage increases, the base voltage also increases. A signal injected at the emitter is used to control transistor conduction in this manner. The changing emitter voltage due to the application of the signal produce changes in base voltage. This produces changes in the conduction and the resistance of the emitter–collector terminals of the transistor.

Signals applied to the emitter decrease conduction as they rise in value. This produces a decrease in current flow. It also produces a larger

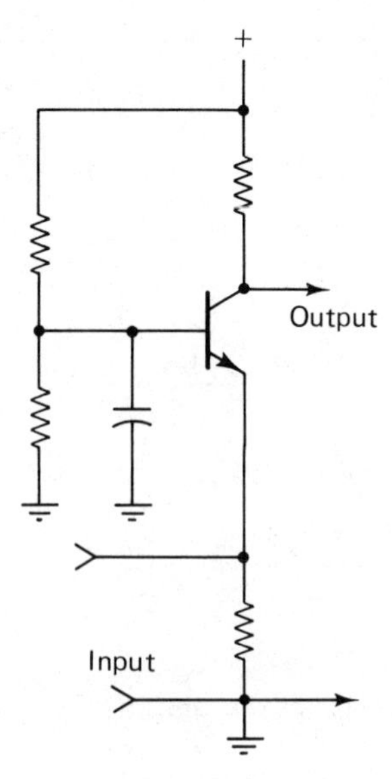

Figure 3-10 Schematic diagram for a common-base amplifier circuit. Signal is injected between the emitter and base and removed between the collector and base.

voltage drop across the output terminals of the transistor. The input and output signals remain in phase in this circuit. The voltage at the base that is required to produce these changes is relatively small compared to the voltage that develops at the output terminals of the circuit. This leads to a large voltage-gain factor.

The development of the transistor integrated circuit (IC) has led to one other type of amplifier circuit. This is shown in Figure 3-11. One has no way of knowing the gain or the phase relationship of input or output signals unless a reference manual is consulted. Each major manufacturer of ICs has a reference manual that provides this information. Most of the references provide circuit descriptions, applications, and operating data. This information is essential if one is to service ICs.

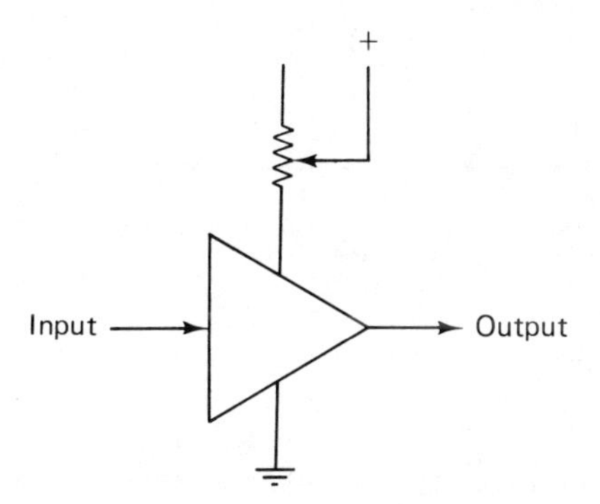

Figure 3-11 An IC amplifier circuit does not indicate the internal wiring of the system.

The material presented thus far in the chapter covers how the transistor amplifier functions. Let us now examine the classification system used for amplifiers. A system that is related to the bias point of the input circuit is used for this classification. The bias point of the amplifier relates to the duty cycle. This is developed when one uses the operating point of the transistor characteristics. The information shown in Figure 3-12 is used for reference. This shows the operating characteristics of the collector circuit. The curves show the amount of collector voltage and current for a given base-current bias. Each of the curves is developed by applying a fixed bias to the circuit and then increasing the source voltage. As source voltage increases, so does the collector current. These two values are used to develop each of the curves shown in the illustration.

The resulting curves show that when the applied source voltage is at zero, there is no current flow in the circuit. Current rises rapidly at first. It then continues to increase until the collector voltage approaches the value of source voltage. The current then levels off and does not increase. This condition is called *saturation*. The condition of no current flow is called *cutoff*. Very few amplifier circuits are operated in the saturated mode. Keep in mind that once saturation is reached, the collector voltage has very little to do with the collector current.

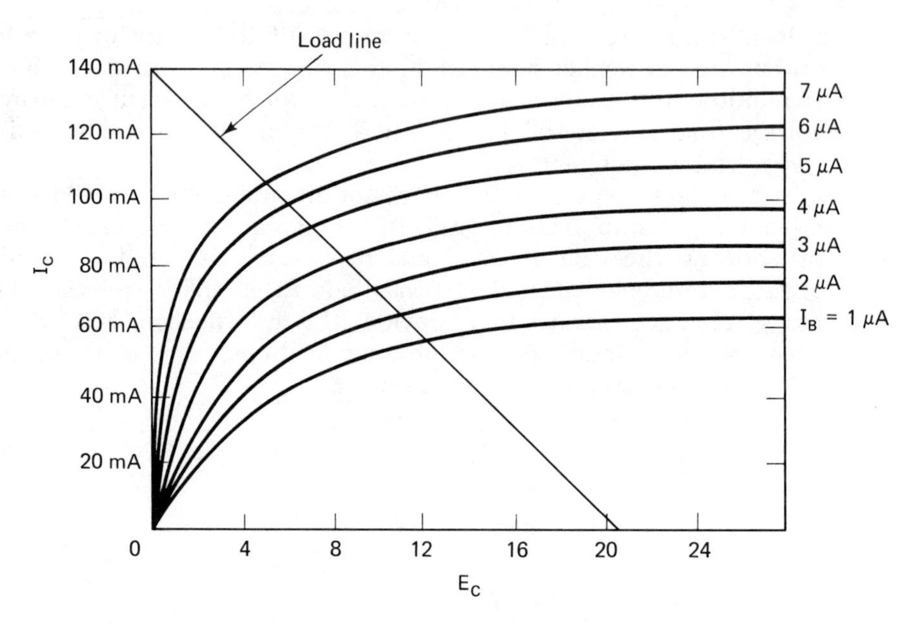

Figure 3-12 Operating characteristics of the collector circuit of an amplifier.

Further examination of the graph shows that a small change in base current controls a larger change in collector current. This, of course, is typical for an amplifier. A fixed bias applied to the base will develop an *operating point* for the circuit. This point is at the intersection of the collector curve and a diagonal line drawn on the graph called the *load line.* The load line shows the different voltage and current values that are established for specific bias values. To find these values, all that is necessary is to draw a horizontal line from the operating point to the vertical leg of the graph. This indicates the operating current for the circuit. A vertical line from the operating point will intersect the voltage leg of the graph. This indicates the collector voltage for these specific operating conditions. Moving the operating point will change the operating conditions. These are found in the same manner as the points described previously. The use of the load line shows the operating conditions of the amplifier circuit. This principle is shown in Figure 3-13. The signal causes the operating point of the amplifier to shift along its load line. This establishes collector voltage and current values. It is possible to design an amplifier whose operating point falls anywhere on the load line. The relation of the no-signal operating point to the load line is used to classify amplifiers. There are three basic classifications for amplifiers that have been utilized for many years. In addition, there are two new classes. Each of these relates to the operating point and duty cycle of the amplifier.

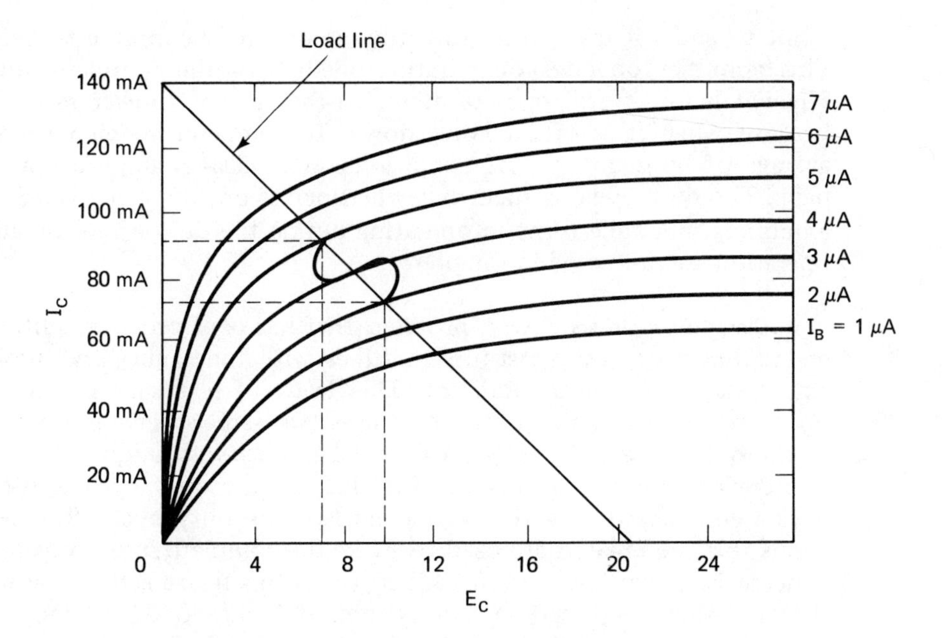

Figure 3-13 Load-line analysis of an amplifier showing circuit conditions.

Class A operation. In this class the amplifier is biased so that it is always on. The swing in the operating point does not allow the amplifier to go into either cutoff or saturation. The usual place for the operating point is at the midpoint of the load line. Class A amplifiers are used when a faithful reproduction of the signal is required. Class A amplifiers are mostly medium- to low-power systems. The transistors do not develop much heat as they operate.

Heat is a problem with all electronic devices. During operation a certain amount of heat will develop. This is due to the resistances in the circuit to current flow. The amount of heat that develops in a tube or transistor is called its *power dissipation.* Each conductive device has a power dissipation rating in units of the watt. When the device is operated for any period of time beyond this rating, it will fail. Most high-power semiconductor devices are operated in conjunction with a heat-radiating device called a *heat sink.* This method of cooling is discussed in Chapter 10.

Class B amplifiers. When design demands power that exceeds the limits of a single amplifying device, it is possible to operate two or more amplifiers in one circuit. Each amplifying device then has a reduced duty cycle. The duty-cycle time is determined by the operating bias point of the circuit. A class B amplifier is biased at cutoff. When the signal causes the operating point to shift, it may either turn the circuit on or continue to keep it off. This amplifier operates only during one-half of the complete duty cycle. When it is on, it will operate at higher-than-normal power ratings. This is because when it is off, it cools down. Its temperature then becomes the average of both on and off conditions. Most class B amplifiers are used in pairs. The duty cycle is such that when one is on, the other is off. The two amplifiers thus take turns in operating during the duty cycle. Details of this operation are presented in Chapter 4.

Class AB amplifiers. One problem that has confronted amplifier designers is that very few transistors or tubes with the same part number have exactly the same characteristics. This leads to problems in designing the operating cycle of an amplifier that uses two same-style transistors or tubes in the output stage. These stages are called *push-pull* output stages because one device conducts during the first half of the duty cycle and the other device conducts during the second half of the duty cycle. The two waveforms that are created are used to make the complete signal waveform. This concept is shown in Figure 3-14. Part (a) of this figure is the ideal waveform that is being developed in the system. It is developed, as shown in part (b), by using two halves of the signal. Under ideal conditions these two signals align at the 180°, or half-wave point.

Figure 3-14 The ideal output wave (a) as desired from a push-pull amplifier is developed from the conduction of each transistor (b).

In practice, the two signal halves often fail to meet as they are designed. This leads to the condition shown in Figure 3-15(a). There is a distortion of the waveform at the point where the two waves cross. This is called *crossover distortion* of the wave. To adjust for this condition, the amplifiers are adjusted so that each is on for more than half of the duty cycle. This provides for a more desirable crossover, as shown in part (b) of the figure. Amplifiers that are biased so that they are on for more than one-half of their duty cycle are classified as class AB amplifiers. Many audio power output stages of amplifiers operate as class AB amplifiers.

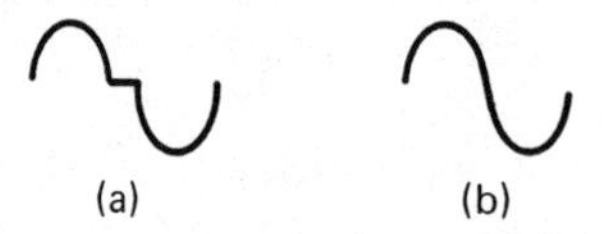

Figure 3-15 Crossover distortion (a) is minimized by applying the correct bias to the base circuit; the result is shown in (b).

Class C amplifiers. The fourth general class of amplifier is the class C amplifier. This amplifier's operating point is well past the point of cutoff. This amplifier is used when very high power is required. It operates for less than half of its duty cycle. Energy that develops during the operating cycle is stored in a resonant circuit. When the operating cycle is over, this energy is used to complete the duty cycle. Almost all applications of class C amplifiers are found in RF circuits.

Other classifications. Three additional classifications for amplifiers have been introduced recently. These are classes D, G, and H. These systems attempt to overcome the high heating effect of a power amplifier. The class G and H systems use two sets of power amplifier transistors in the output stage. In the class G system, one set operates as a low-powered output. It handles most of the signals. The second set of output transistors operates only on the audio power peaks. Use of this two-set output system permits high power when it is required and also permits the system to operate with a lower power for those frequencies that will reproduce at the lower power rate.

The advantage of the system described in the preceding paragraph is that it does not require heavy power supplies or large heat sinks for the output stages. The disadvantage of this system is that it cannot deliver high power for extended periods of time. The class H system uses a larger power supply for the second set of power output transistors. It is able to deliver high power output over extended periods of time. It does require a heavy and large power supply for the amplifier.

The class D amplifier uses a different principle. It uses a square-wave signal in its output stage. The square wave requires very little power. The square wave is used as a carrier wave for an audio signal. The audio signal modulates the square wave in a manner similar to that of the broadcast signal of a radio station. The carrier is used as a means of delivering signal power to the speakers. This system has one severe disadvantage. It is that the square-wave system is full of harmonics. These harmonics radiate well into the megahertz range. They cause RF interference in other systems. The class D amplifier requires heavy shielding and filtering to overcome this deficiency.

Material in this chapter covered the theory of amplifiers. It also covered the classification systems used and how they are determined. Material in the next chapter discusses the various output systems in use and how signals are processed through amplifier systems.

QUESTIONS

3-1. Define the term "amplifier."

3-2. Describe the amounts of current flow in the major paths of the transistor amplifier.

3-3. What are the three "common" classifications of amplifiers?

3-4. Identify the signal input and output connections for each of the amplifiers identified in Question 3-3.

3-5. What is the transistor condition when it is in cutoff?

3-6. What is the transistor condition when it is in saturation?

3-7. Name the classes of amplifiers that are based on the conduction conditions.

3-8. When are each of the classifications of Question 3-7 used in an amplifier?

3-9. How does the signal output voltage of an amplifier develop?

3-10. Explain how an input signal is used to control the output circuit of the amplifier.

Chapter 4

Transistor Amplifier Systems

Successful service technicians must have a good understanding of the theories of operation of basic circuits. This knowledge is necessary in order to evaluate the circuit to determine if it is working properly. This evaluation includes knowing how the circuit is supposed to function, where to make evaluation tests, and what to expect to find when making the test. Few, if any, manufacturers of stereo equipment provide waveform information with their service literature. Technicians are expected to know what to expect to find as they troubleshoot a stereo system.

The only way possible to learn about stereo systems and the approximate waveforms is to understand the basic building blocks used for the system. One should be able to recognize a basic amplifier circuit. Since there are only three of these, it is not too difficult. In addition, one should know the basic waveform phase relationship between input and output circuits. The basic circuits were described in Chapter 3. This chapter explains the system used for audio amplifiers. Individual blocks are put together in order to form a total system. Signal transfer between blocks, or stages, is discussed. In addition, output and driver stage basic theory is also presented. Once the theories of amplification are known, the application of troubleshooting rules can be discussed. The material related to troubleshooting amplifiers is discussed in Chapters 10 and 11.

SIGNAL TRANSFER METHODS

There are three systems used to transfer from one amplifier stage to another amplifier stage. The process of transferring signals is called *signal coupling.*

The three systems are termed direct coupling, resistor–capacitor (*RC*) coupling, and transformer coupling. Each has advantages as well as disadvantages. An understanding of how each functions is required information for the service technician.

Direct coupling. A schematic for a direct-coupled amplifier is shown in Figure 4-1. This circuit uses two transistors. One of the transistors is a PNP and the other is an NPN type. The waveforms shown are typical for the amplifier configuration. Both stages are common-emitter circuits. This develops a phase reversal between input and output and a voltage gain. Signal flow in this system is across the circuit in a horizontal path. The input element of transistor Q_1 is its base. The output element is its collector. The load resistor for this transistor is R_2. The variations in dc operating voltage that develop at the collector due to transistor action are identified as the signal.

The process of developing this signal is as follows. A positive-going signal at the input to the circuit causes the transistor to turn on by increasing the forward bias on it. This produces an increase in current flow in the emitter-collector circuit and a voltage drop at the collector. Thus, the collector voltage decreases as the base voltage increases. When the input signal decreases, the current in the transistor also decreases. This produces an increase in collector voltage. The swing in collector voltage is greater than

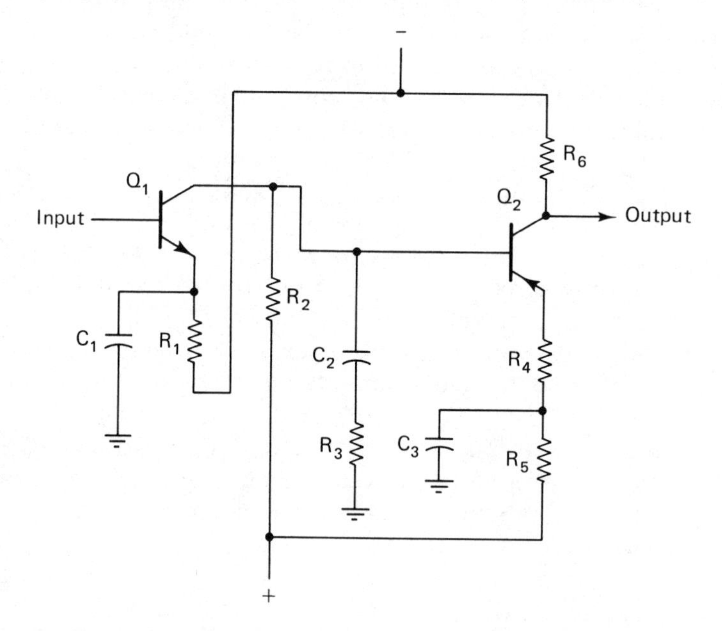

Figure 4-1 A direct-coupled amplifier schematic shows no coupling device between stages.

the swing in base voltage that caused the changes. This is the process of amplification.

Transistor Q_2 is a PNP type. Its polarity is opposite to that of the NPN transistor Q_1. The base of Q_2 transistor is wired directly to the collector of Q_1. Thus, any changes that occur at the collector of Q_1 directly affect the operation of the second transistor. The signal waveform at the base of Q_2 is opposite in polarity to that at the base of Q_1. Transistor Q_2 increases its conduction as the base bias decreases. This is opposite to the action of transistor Q_1. A decreasing base signal will produce an increase in conduction in the second transistor. As the base signal increases, the conduction through Q_2 will decrease. This produces an amplified and opposite polarity signal at the output of Q_2.

The factors that control the amount of amplification of any amplifier relate to the gain of the transistor. This refers to the design of the device. It indicates the amount of amplification one can expect when the device is used as an amplifier. The swing in collector voltage of Q_1 in this circuit is applied directly to the base of Q_2. The second transistor must be able to have its collector swing in a similar manner. If the swing due to base-signal application drives the second transistor into either cutoff or saturation, signal distortion occurs. Distortion does not permit faithful reproduction of the signal. This is undesirable in a class A amplifier. The advantages of a direct-coupled amplifier are that it has an almost unlimited frequency response and that it uses fewer component parts than other types of amplifiers. The major disadvantage is that when one component fails, its failure disturbs the entire system. Repair of this system is discussed in Chapter 11.

RC coupling. A second method of passing signals from one stage to another is called resistor–capacitor (*RC*) coupling. A schematic for a *RC*-coupled amplifier is shown in Figure 4-2. A network consisting of a resistor and a capacitor is used as a signal-coupling device. The capacitor in the network permits only the voltage changes that occur at its input plate, or

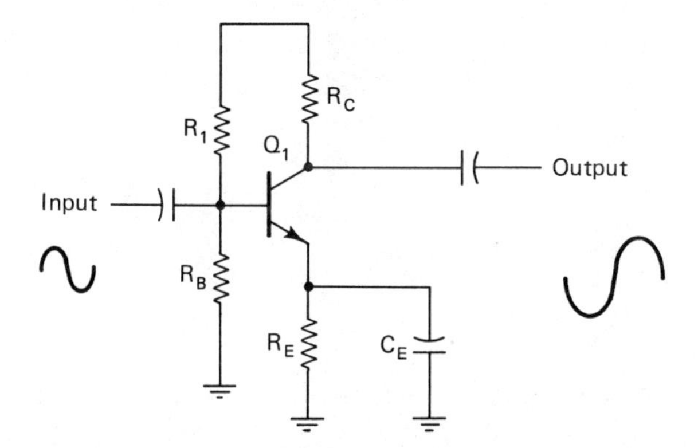

Figure 4-2 An *RC*-coupled amplifier circuit uses a resistor and a capacitor for signal coupling.

left side in the schematic, to be transferred to its other plate. This is also true for the capacitor used to transfer signal from Q_1 collector to Q_2 base.

The dc-coupled amplifier collector voltage swing is passed on to the next stage. In the RC-coupled amplifier only the variation in dc voltage swing is passed on to the input of the next stage. This signal voltage is usually lower in voltage value than the operating voltage found at that point. The capacitor is charged and discharged through the base-bias resistors so that only the changes in voltage that occur during amplification are transferred to the next stage. These coupling capacitors effectively isolate one stage from another. A component failure in one stage does not affect any other stages, as is found in the dc-coupled amplifier.

Transformer coupling. The third method of transferring signals between stages is called transformer coupling. A schematic diagram for this type of circuit is shown in Figure 4-3. In this circuit one transformer is used as the input and another as the output coupling device. The input signal causes a current flow in the primary of transformer T_1. This produces a current flow in the secondary of the transformer. This secondary current is used to develop base swing. The changing base current produces the output signal seen at the collector of Q_1.

Transformer T_2 is the output transformer for this circuit. It is also the load for the collector circuit. Current flow variations through the transformer develop voltage variations at the secondary of the transformer. The secondary is connected to the next stage. It is connected to a complete circuit and current flows in the secondary due to the changing voltage. In this manner the signal changes are passed from one stage to another by transformer action. The specific size of the waveforms at the secondaries of each transformer depends on the turns ratio of the device.

The application of the transformer is shown in Figure 4-4. In effect, the transformer is an impedance-matching device. Its purpose is to transfer the maximum amount of power from the output circuit of one stage to the input circuit of the next stage. Each of its windings is designed so that

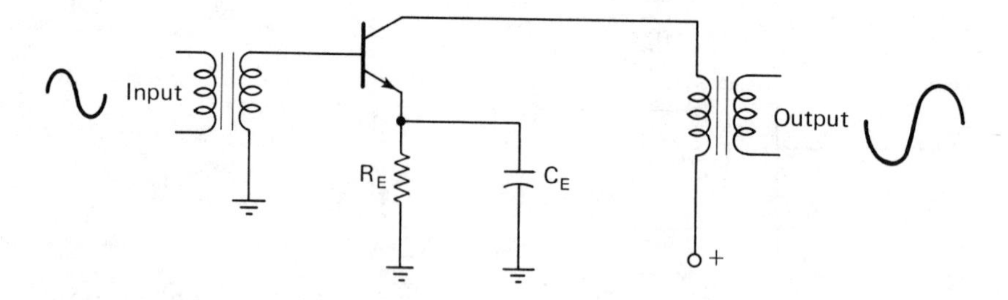

Figure 4-3 Transformer coupling between amplifier stages.

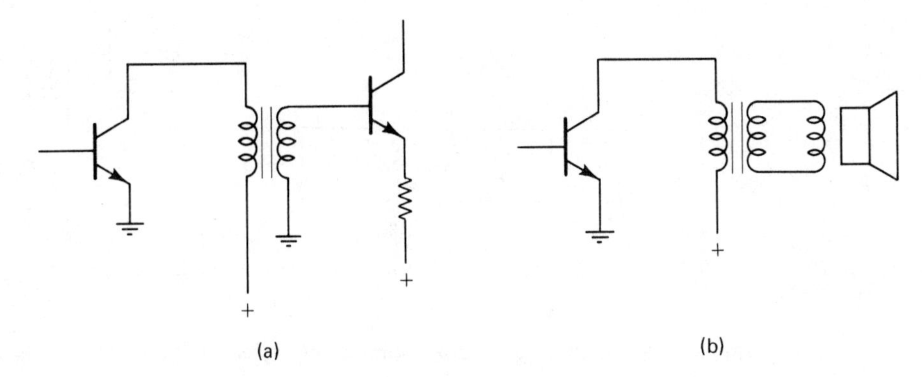

Figure 4-4 Transformers are used as impedance-matching devices in these circuits.

this can be successfully accomplished. Part (a) of the drawing shows signal transfer between stages. Part (b) of the drawing shows signal transfer from the final, or output, stage to a speaker. Power-handling requirements as well as correct impedances are important considerations in each of these circuits.

The advantage of using transformer coupling is that these circuits will handle more power and higher gain than either *RC* or coupled circuits. The principal disadvantage of using a transformer is related to size and weight of the device. A second disadvantage is that the cost of a circuit using transformers is greater than that of a circuit using the other coupling systems.

OUTPUT AND DRIVER STAGES

Almost all output stages used in audio equipment use a two transistor push-pull type of circuit. There are several variations of the basic push-pull circuit. These include quasi-complementary, complementary, and transformerless circuits. Each is discussed in the following section. Signals have to be developed in both the amplitude and phase in order to provide the necessary drive to the final stage. Very often this development is done in a stage called the *driver.* A block diagram for this type of system is shown in Figure 4-5. Two or more voltage amplifiers are used in order to develop sufficient signal strength. This signal is then fed to a driver block. The output of this block depends on the requirements of the output stage. It may be two signals, each 180° out of phase with the other. It may also be a single signal. Each schematic will show the specific circuit. Keep in mind that although the discussion here is related to single-channel operation, the stereo system uses two identical amplifier systems.

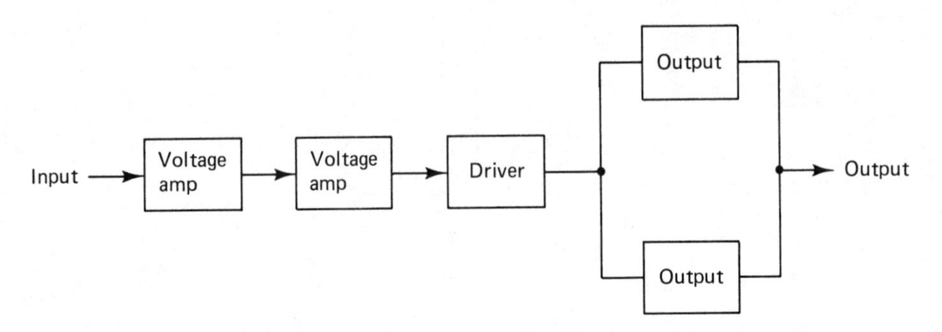

Figure 4-5 Block diagram for an amplifier system using a driver stage.

OUTPUT STAGES

Let us start our coverage of driver and output stages by first examining the various output-stage circuits. After these circuits are explained, we can look at the driver stage and see how it is used to meet the requirements of the output stage. The circuits used as examples are simplified versions of the actual working circuits. This is done so that the basic concept of performance is understood.

Push-pull circuits. Early developments for two-device audio output stages used this arrangement. The reason for this is that the circuits were designed and patterned after vacuum-tube circuits. There is no such thing as a NPN and a PNP vacuum tube. Conduction is only one way through the tube. Circuits had to be designed that shifted the phase of the input signal in order to use a class B or AB amplifier. The circuit shown in Figure 4-6 is a transistor version of this type of system. Two identical transistors are employed. In this circuit both are NPN. They also could be a pair of PNP types if the power source connections were reversed.

Each of the output transistors conducts during half of the input wave cycle. This allows operation at higher than normal conditions during the on-time. When off, the transistor cools. In this manner it does not become destroyed due to overrated operation. The transistors are biased at, or near, cutoff. The only time they conduct is when the input signal takes a positive swing. Shifting the phase of the input signal as it is applied to the two transistor bases permits a switching of operation from one transistor to the other.

Current flow for the two output transistors is the same during their conduction periods. The arrows on the schematic show this direction. The primary of the output transformer T_1 is center tapped. The center tap is connected to source positive. Current flow during the conduction cycle of the two transistors is in an opposite direction through the primary of this

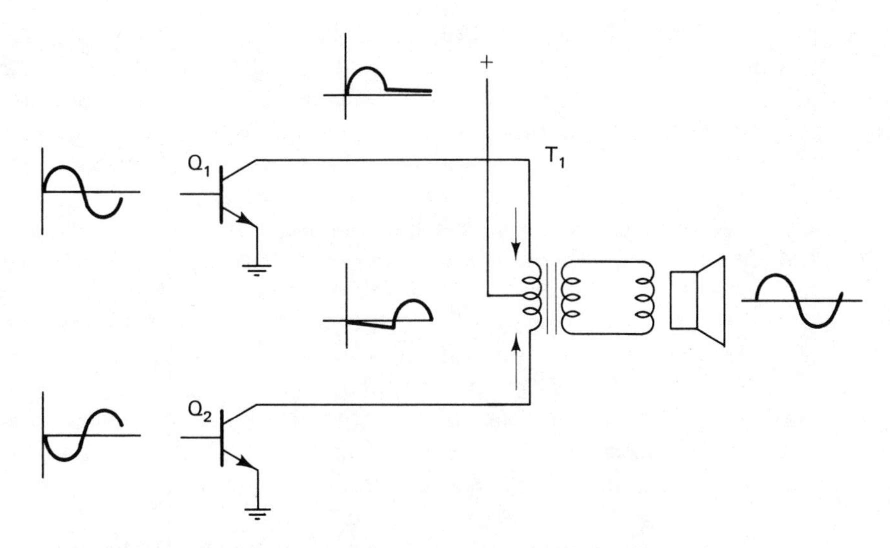

Figure 4-6 Push-pull audio output stage using identical NPN transistors.

transformer. The waveforms shown at the collector of each transistor are the result of the current flow. Both waveforms are positive going in form. They do show a phase shift. This phase shift causes a current flow through the primary of the output transformer in two directions. The result of this action is a full waveform at the secondary of the transformer. In this manner the signal is separated into two halves, amplified, and then put back together in order to drive the output speaker.

Output transformerless system (OTL). The audio output transformer takes a lot of space. In addition, it is heavy and costly. There are circuits that do away with this transformer. One such circuit is shown in Figure 4-7. This is called an output transformerless system (OTL). In the circuit shown, two same-polarity transistors are series connected between common and the

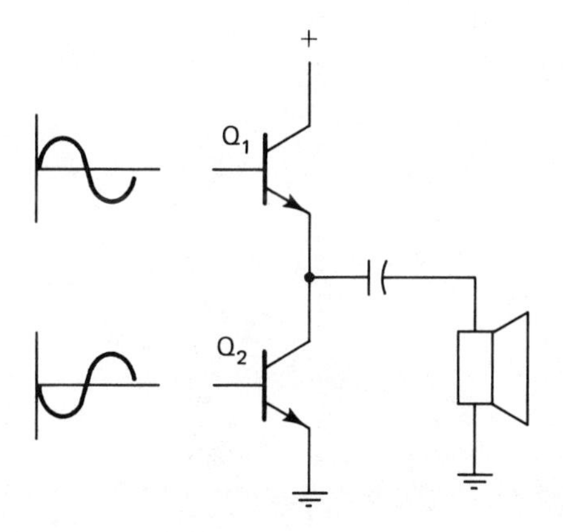

Figure 4-7 Transformerless push-pull audio output stage using identical transistors. Coupling is accomplished by connecting a capacitor to the midpoint in the circuit.

power source. The output connection is at the midpoint between the two transistors. The output coupling device is a capacitor. This capacitor is series connected to a speaker. Two out-of-phase input signals are required to make this circuit function. Both transistors, being NPN types, require a positive-going signal in order to turn them on. With no signal applied, the voltage at the midpoint of the transistors is equal to one-half of the source voltage. When Q_1 is turned on, its emitter–collector resistance decreases. This raises the dc operating voltage at its emitter. When Q_2 is turned on during its half of the operating cycle, its emitter–collector resistance also decreases. This produces a drop in the dc voltage at its collector. The Q_1 emitter and Q_2 collector junction is the output point for this circuit.

Voltage variations at the output point develop a varying charge on the plate of the capacitor that is also connected to this point. This, in turn, changes the charge on the other plate of the capacitor. The varying charges transfer energy from the transistor circuits to the speaker.

Complementary-symmetry output. One advantage of the transistor is that we have both NPN and PNP types with identical operating characteristics. This led to the development of complementary-symmetry output circuits, shown in Figure 4-8. With this type of circuit there is no need for a split-phase input signal. Except for operating bias considerations, both input connections are wired to the same point. When the input signal goes positive, it turns on the NPN transistor Q_1. This raises the voltage at its emitter. When the input signal goes negative, it turns on the PNP transistor Q_2. In this circuit both transistor emitters are connected to the midpoint of the circuit. With no signal applied, the dc voltage at this point should be half that of the source. The changes in operating voltage at the emitters of the output transistors is used to develop a charge on the plates of the coupling capacitor. This, in turn, develops a current flow through the speaker circuit, producing sound waves.

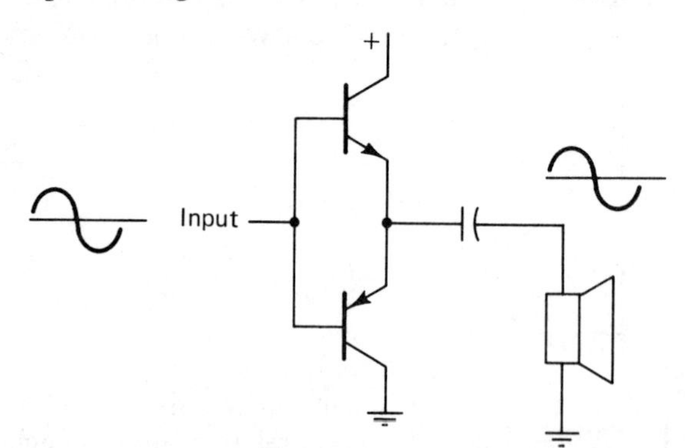

Figure 4-8 A complimentary symmetry audio output stage uses two identical, but opposite polarity, transistors.

Quasi-complementary-symmetry output. This circuit uses a complementary-symmetry system to drive a push-pull amplifier using the same polarity

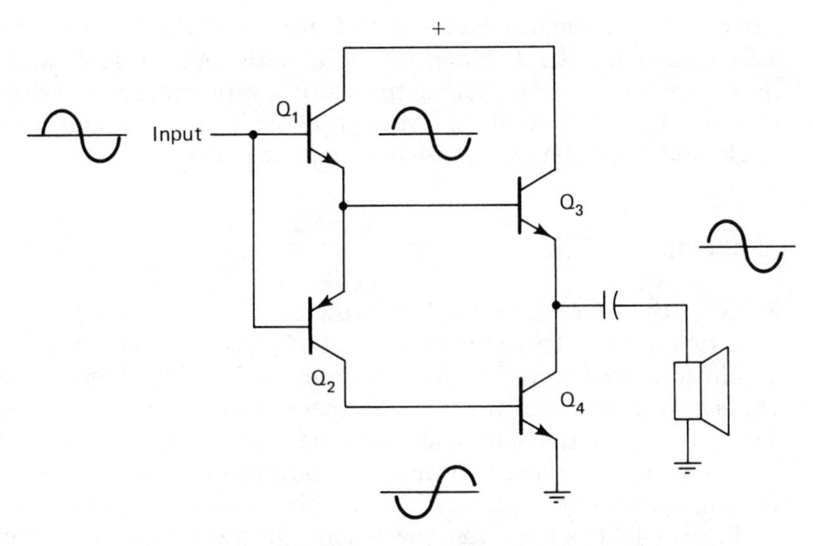

Figure 4-9 Quasi-complementary audio output system.

transistor. A circuit utilizing this concept is shown in Figure 4-9. Transistors Q_1 and Q_2 are connected in a complementary-symmetry circuit. Their output circuits are out of phase with each other. Transistor Q_1 uses an emitter-follower circuit. Transistor Q_2 uses a common-emitter circuit. This develops the two out-of-phase signals required to drive the output transistors Q_3 and Q_4.

The output for this circuit is at the midpoint of the Q_3-Q_4 circuit. Voltage variations at this point are coupled through the capacitor to the speaker. These produce a current flow in the speaker circuit. The use of two pairs of output transistors develops power for the output that is higher than that developed by a single pair of output transistors.

Darlington-pair amplifiers. One method of obtaining a large amount of amplification from a circuit is to connect two transistors in a Darlington

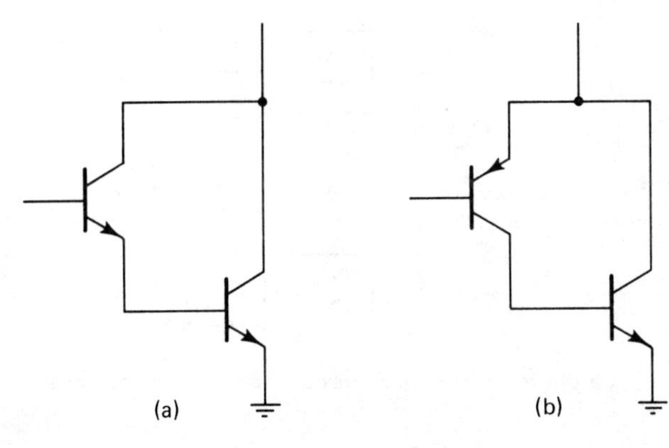

Figure 4-10 A Darlington-pair transistor circuit provides very high gain.

circuit. The gain in any circuit is the product of the gain factor in each stage. This is true for the Darlington configuration. Transistors are connected as shown in Figure 4-10. These transistor configurations develop gain figures that run in the thousands. Darlington-pair transistors may be purchased as single units from several transistor manufacturers.

DRIVER CIRCUITS

Some of the schematics used to show the layout of the power output stage have two inputs. These inputs are out of phase by 180°. A circuit other than a voltage amplifier is required in order to develop a signal with the proper phase and amplitude to drive the final output stage. This circuit is called a *driver.* It takes the audio signal from the voltage amplifier and prepares it so that it can be used to drive the output devices. Some of these circuits are called *phase splitters* because of their action. The block diagram shown in Figure 4-11 shows the placement of the blocks and their respective waveforms.

There are two basic methods used to split the phase of the audio signal. One of these requires a transformer. The other one uses an amplifier stage with output connections at both the emitter and the collector.

Transformer drivers. A transformer is often used to establish a dual output signal. A circuit using a transformer for this purpose is shown in Figure 4-12. The primary of the transformer is connected as a load to the collector circuit of transistor Q_1. The secondary of this transformer is center tapped. The center-tap connection is connected to circuit common. When a changing current flows in the primary, a voltage develops in the secondary of this transformer. There are two voltages that develop because of this action. The center tap is at reference zero volts. The voltage that develops in the secondary has both a positive and a negative value. This is shown by the waveforms in the schematic. These two voltages are 180° out

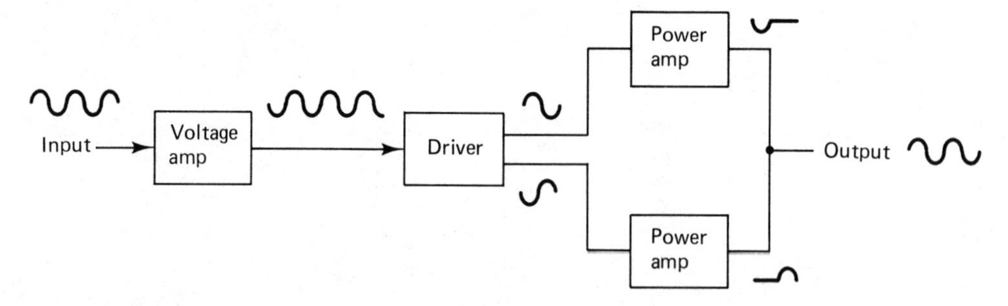

Figure 4-11 Block diagram for a phase splitter. This circuit has two out-of-phase outputs.

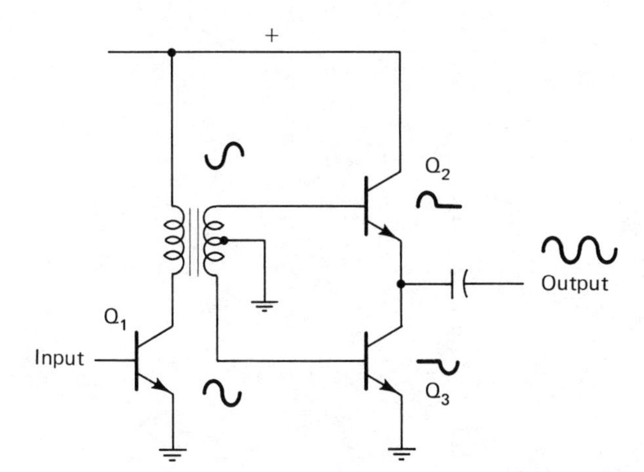

Figure 4-12 A transformer driver/phase splitter circuit provides two out-of-phase signals.

of phase with each other. The action here is very similar to the action of a full-wave center-tapped power transformer and rectifier circuit.

The circuit from the secondary of the driver transformer is used to bias the output transistors so that they are operating in class AB. The result of this develops a push-pull action. This meets the power requirements of the output stages.

Transistor drivers. The circuit shown in Figure 4-13 is a transformerless driver circuit. In this circuit a transistor amplifier is used. This amplifier is

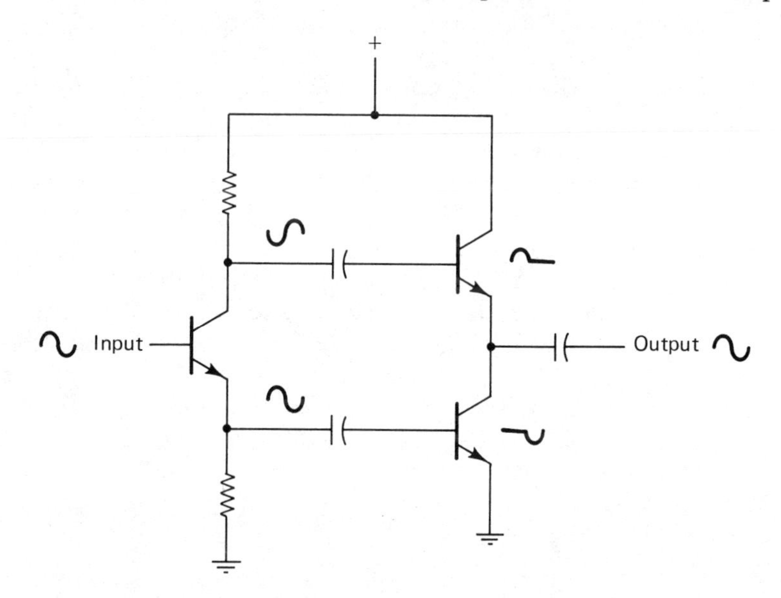

Figure 4-13 A transistor drive/phase splitter circuit also provides two out-of-phase outputs.

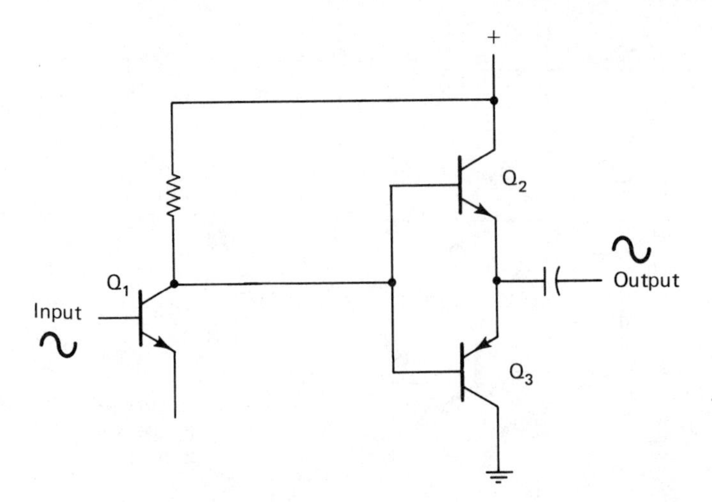

Figure 4-14 A complementary-symmetry output system does not require a phase-splitter stage.

actually a common emitter and a common-collector amplifier. It is called a *transistor phase splitter.* Two output circuits are used. One of these is connected to the collector. It provides an inverted output as is found when using a common-emitter amplifier. The second output is connected to the emitter. This output produces the in-phase signal found with a common-collector amplifier. The size of the resistors in the emitter and the collector determine the amplitude of the two signals. Usually, these are selected to provide equal-amplitude signals at both output points. The signals are used to drive the final output stages.

Either of these two circuits may be used with any of the push-pull amplifier systems. This system is not required for complementary, quasi-complementary, or OTL circuits. These circuits are fed directly from the voltage amplifier. A circuit for this type of system is shown in Figure 4-14. The output of Q_1 is fed directly to both bases of the output transistors Q_2 and Q_3. The base circuit is biased so that the output transistors are operating in class AB. These output transistors are connected in emitter followers. There is no phase inversion for this part of the circuit.

The common signal is used to turn on each transistor during one-half of the signal cycle. In a practical circuit there are resistors and diodes used for establishing the correct operating bias. The conduction of each transistor is used to develop the composite signal shown at the output terminals of the power transistors.

INTEGRATED-CIRCUIT SYSTEMS

The circuits discussed so far use discrete components. These same circuits are now available as integrated circuits. The packaging of these ICs varies but

LM1877 Dual Power Audio Amplifier

General Description

The LM1877 is a monolithic dual power amplifier designed to deliver 2W/channel continuous into 8Ω loads. The LM1877 is designed to operate with a low number of external components, and still provide flexibility for use in stereo phonographs, tape recorders and AM–FM stereo receivers, etc. Each power amplifier is biased from a common internal regulator to provide high power supply rejection, and output Q point centering. The LM1877 is internally compensated for all gains greater than 10.

Features

- 2W/channel
- −65 dB ripple rejection, output referred
- −65 dB channel separation, output referred
- Wide supply range, 6–24V
- Very low cross-over distortion
- Low audio band noise
- Internal current limiting, short circuit protection
- Internal thermal shutdown

Applications

- Multi-channel audio systems
- Stereo phonographs
- Tape recorders and players
- AM–FM radio receivers
- Servo amplifiers
- Intercom systems
- Automotive products

Connection Diagram

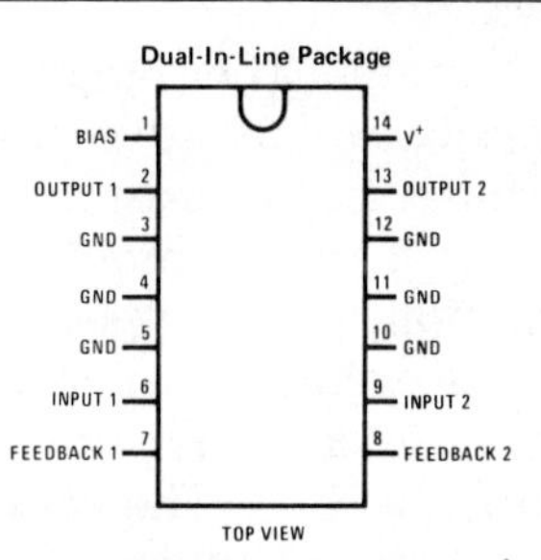

Order Number LM1877N
See NS Package N14A

Equivalent Schematic Diagram

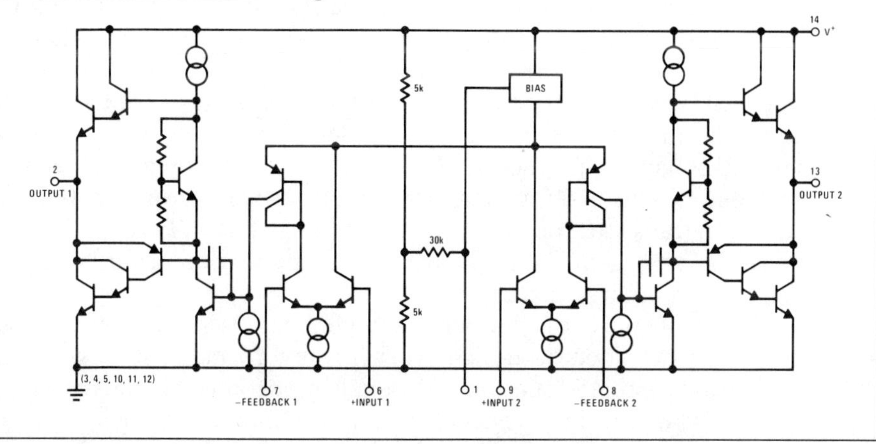

Figure 4-15 IC stereo amplifier contained on one chip. (Courtesy National Semiconductor Corporation.)

the function is the same. IC amplifiers are available with power output ranges of from 0.5 W to well over 25 W. Several IC units have two separate amplifier systems in the same package. Technical information relating to specific systems is available from both the IC manufacturers and replacement IC sources. All that the technician requires is the values of signal and operating voltages and the wiring connections from the internal electronics to the pins on the outside of the package. An IC stereo amplifier and its schematic diagram are shown in Figure 4-15. The only components that are not included inside the IC are the coupling and bypass capacitors. All the other components are included in the package.

V-MOS TECHNOLOGY

True sound reproduction is difficult to achieve. Efforts have been made to improve the quality of the sound system. Many arguments have developed as to whether the vacuum tube or the transistor is the better system to use. Each side has its advocates. Their arguments are valid. The position of this author is that perhaps both sides are correct.

Both vacuum tubes and transistors have the ability to act as amplifiers. They also introduce a certain amount of distortion to the signal as it is reproduced. In addition, both types of devices have specific frequency limitations. Efforts to overcome these limitations have led to the development of a new type of semiconductor power device. It is called a vertical field-effect transistor, (VFET). The VFET has several advantages over the junction transistor. These advantages include the following. The VFET has excellent frequency response. This provides a device that has a very low amount of frequency distortion. It can be produced as either a N- or a P-type semiconductor. This allows the application of the VFET in complementary-symmetry circuits. The device also has a high input impedance and a low output impedance. This makes it ideal for use as a power output transistor. One last advantage is that the VFET is a voltage-controlled device. The junction transistor is current controlled. This type of control can lead to thermal-runaway problems. The vertical FET's operation overcomes this problem. It also eliminates the circuit design problems that occur with junction transistor power output circuits. The VFET is a new device. It certainly will gain in popularity and application as more amplifiers are designed to use its circuitry.

One purpose of the service technician is to repair a defective unit. This does not include design work in any form. It is important to understand how systems function. Of equal importance is the knowledge of what to expect to find in a working circuit. The material presented in this chapter will aid in this understanding. Apply this information to schematic diagrams for various audio systems. Attempt to identify the type of amplifier used in each circuit. Identify the phase relationship of the output signal as it compares to the input signal. Also look for signal amplification (or a lack of it) for each stage. Those persons wishing to attempt some "hands-on" learning may wish to use an oscilloscope in order to follow the signal path in the system. Use of the oscilloscope will help to verify the theories presented in this book.

QUESTIONS

4-1. Name the types of signal transfer methods used for amplifiers.
4-2. What is the purpose of a driver stage?
4-3. What is the phase relationship of the output signals of the driver circuit?
4-4. What is meant by the term "push-pull"?
4-5. What is an OTL system?
4-6. What types of transistors are used in complementary-symmetry amplifiers?
4-7. What is a Darlington-pair amplifier?
4-8. How does a phase splitter operate?
4-9. What is the advantage of a VFET over a junction transistor?
4-10. What components are *not* included inside the IC amplifier?

Chapter 5

Input and Output Devices

All stereo units have at least one major common function. This function is to process an electronic signal. The signal is presented to the input terminals of the unit. The input signal at this point is developed from several different types of equipment. The form of the input signal is that of a varying voltage. The varying signal voltage is produced by several different kinds of input transducers. The transducer is a device that changes one form of energy into another form of energy. In the case of stereo equipment sound, radio waves, magnetic fields, and motion are changed into an electrical signal voltage. Material presented in this chapter shows how this change occurs. The first part of the chapter is devoted to a discussion of input transducers. The latter portion covers output transducers and how they produce sound or magnetic waves.

INPUT DEVICES

There are several types of input devices used with stereo equipment. These include tape heads, radio antennas, microphones, and phonograph styli or cartridges. Each input device has its own specific set of characteristics. In addition, each may have certain output characteristics that the others may not have. As the material related to specific devices is presented, keep in mind that they all have a similar purpose. This purpose is to convert the input information into a usable electronic signal. This signal is in the form of a varying voltage. In many cases it has a very low amplitude. The signal requires a great deal of voltage amplification in order to be processed by the amplifier system, hence the low-level amplifier or preamplifier.

The service technician needs to know the system being serviced. Some of the input devices described here have stereo outputs. This means that the output of the device is separated into two channels of information. These are identified as the right channel and left channel. Devices that are capable of producing a two-channel output include tape heads, FM stereo tuners, and phono cartridges. AM tuners, FM monaural tuners, and microphones usually have single-channel outputs. The output connections for these units are wired in parallel when connected to the input of a stereo system. The stereo amplifier output is the same for both channels under these conditions.

A block diagram for a composite stereo system is shown in Figure 5-1. A two-channel audio system is shown. Both stereo and monaural inputs are shown. The switch, S-1, is the function selector switch. It is used to select a specific input to the audio system. This switch has two sections. The section that is connected to the right channel is identified as S_1-1. The other section is used for the left channel. It is identified as S_1-2. Both sections use the S_1 identifier. The number that is used after the subscript 1 refers to the specific section of the switch. Devices that have two separate outputs, such as the tape, phono, and FM stereo, have cables connected to each switch. The monaural devices have their output connections parallel wired to both sections of the switch. Schematic or block diagram for specific units show specifically how the unit is wired. These, and the information provided in this chapter, will help in the service of the stereo system.

Tape players. Tape playback units convert magnetic energy into electrical energy. The art of recording on tape has developed greatly in the past few years. The basic system is shown in Figure 5-2. Two requirements for this system are a recording head and a tape that is capable of storing magnetic information. This drawing of the tape head shows that it has two main parts. One of these is a coil of wire. This coil is called the *armature.* The armature is wound on a magnetic core. The core is used to direct the magnetic lines of force. An electronic signal is used to develop this magnetic field during the recording process. Variations in the strength of the signal develop a varying magnetic field in the head.

The recording tape is made of several layers. These are shown in Figure 5-3. These include the base, a layer of adhesive, and the metal oxide. Several different oxides are in use today. The type of oxide used helps to determine the frequency response of the tape. Early tapes used an iron oxide material. It has limitations related to the upper-frequency limits for recorded material. Later developments include several types of oxides that have other materials, such as chromium, added to them. These additional materials helped improve the high-frequency response of the tape.

At one point in time, during the early days of tape recorder develop-

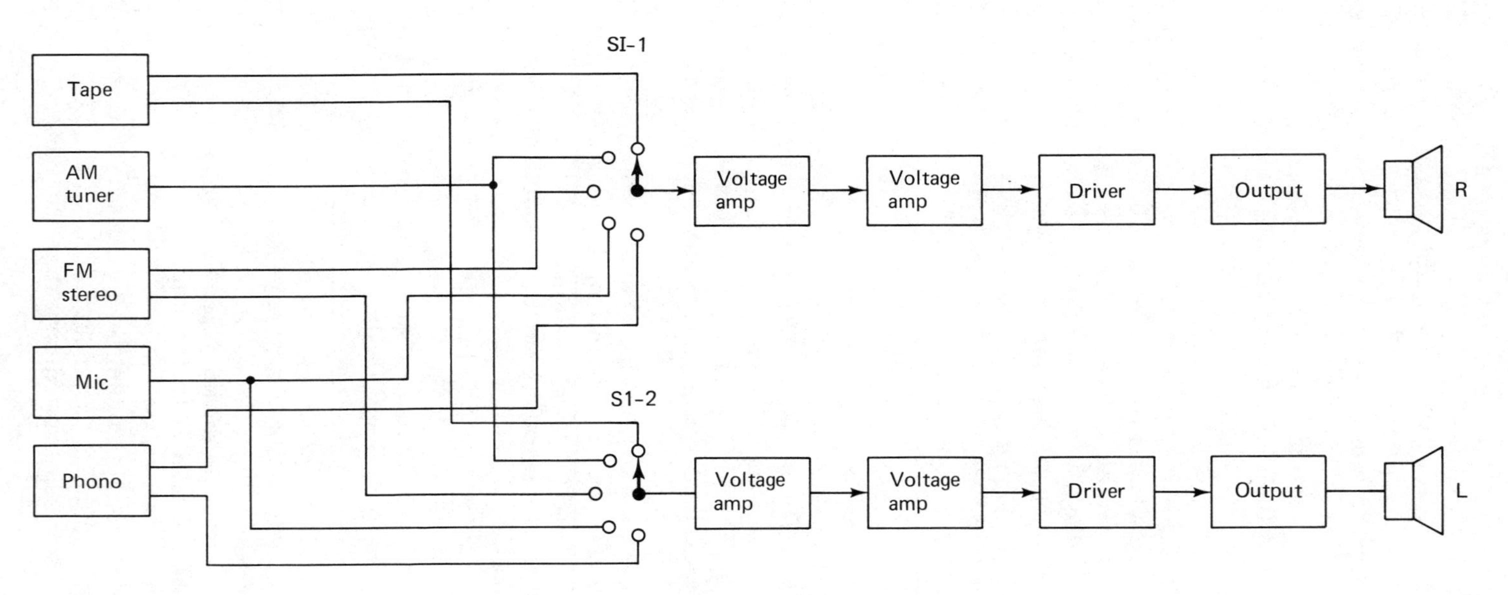

Figure 5-1 Block diagram for a composite stereo system.

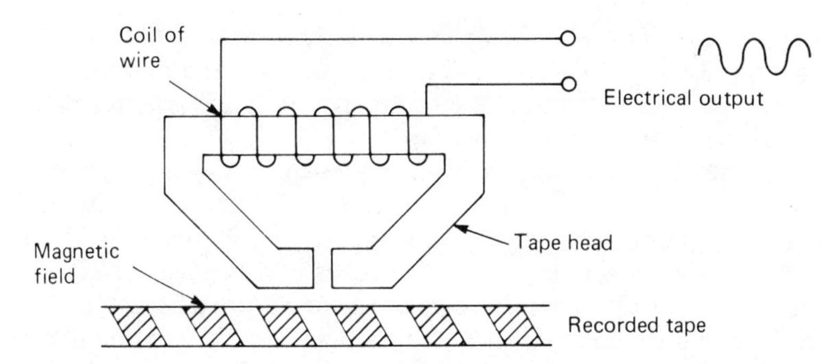

Figure 5-2 Basic tape recording system. (From Joel Goldberg, *Radio, Television, and Sound System Repair: An Introduction*, © 1978, page 32. Reprinted by permission of Prentice-Hall, Inc.)

ment, tape speed was directly proportional to the upper-frequency limitation of the tape. Higher speeds meant high-frequency response. The new oxide mixtures change all of this. Slower tape speed can now produce full-fidelity sound. In many instances the frequency range of the newer materials is well beyond the audio-frequency spectrum.

The magnetic field that is developed in the tape head is used to produce the recorded tape. The tape is pulled past the head by a tape transport

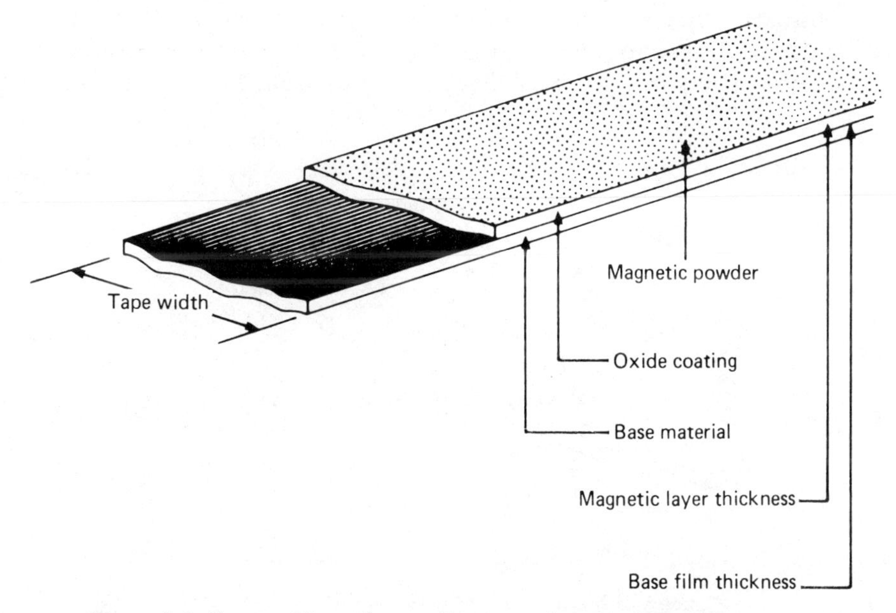

Figure 5-3 Composition of recording tape showing base, adhesive, and oxide layers. (From Joel Goldberg, *Radio, Television, and Sound System Repair: An Introduction*, © 1978, page 32. Reprinted by permission of Prentice-Hall, Inc.)

mechanism. Magnetic fields from the tape head interact with the iron particles on the tape. These particles are arranged in a manner that represents the information used to develop the magnetic field. The magnetic information is then stored on the recording tape.

The reverse of this process is used to play the recorded information on the tape. The principle of electromagnetic induction is used for this process. Each recorded area on the tape has its own magnetic field. The magnetic field, when passed by a wire, induces a voltage on that wire. When several turns of wire are used, such as in the armature of the tape head, the induced voltage is stronger than it is for a single wire. The laminated core of the tape head directs the lines of force to the armature winding. A pair of wires lead from the armature to the first electronic stage of the stereo system. This is shown in Figure 5-4. The voltage that is induced in the tape head has the exact form of the signal voltage used to develop the recorded tape.

A stereo tape has pairs of recorded tracks on it. Each track represents a recorded channel. One track is used for each channel of the stereo system. The exact number of pairs of tracks depends on the system used for recording. There are certain standards for prerecorded tapes. A reel-to-reel system has two to four tracks. Cassette systems have four tracks. Cartridge systems have eight tracks of recorded information. The position of the playback head in relation to the tracks on the tape determines which pair of tracks is being used. Some systems change the position of the head in order to use a different set of tracks. Other systems require that the tape be taken off the transport mechanism and replaced in reverse position. Stereo tape systems require two tape heads for playback. In many systems both heads

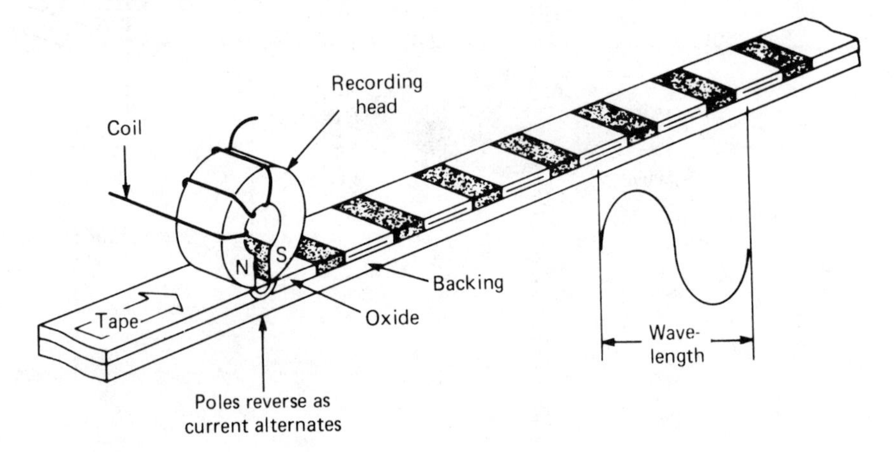

Figure 5-4 Method of inducing a signal from the tape head onto the tape. (From Joel Goldberg, *Radio, Television, and Sound System Repair: An Introduction*, © 1978, page 14. Reprinted by permission of Prentice-Hall, Inc.)

are housed in one metal boxlike container. This is shown in Figure 5-5. At the rear of the housing are four connectors. These are for the two heads. Each head uses one pair of connectors. At the front are two dark, rectangular boxes. These represent the part of the tape heads that are used for pick-up of the magnetic information from the tape.

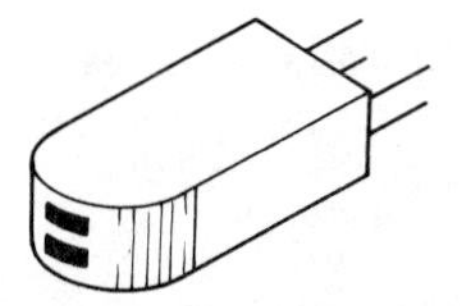

Figure 5-5 A tape head for a stereo playback system has two heads in one container.

Some tape recorder and playback units use the same heads for both recording and playback. In addition, erase heads are often built into the same housing. These appear as two additional black rectangular boxes. Service literature for a specific unit is used to identify the number and type of heads found in this housing.

Antennas. Another important input transducer is the antenna. Signals are created by the broadcast station. These signals are electromagnetic in form. The length of each wave is a function of the frequency of the signal. The electromagnetic waves will induce a voltage in a wire as they cross it. If the waves cross the antenna at an angle of 90°, the maximum signal is induced onto the wire. The signal diminishes in strength as the distance from the transmitter to the receiving antenna increases. Received signals usually have a value that is in the microvolt range. Since the signals are higher in frequency than audio waves, they cannot be processed by the audio amplifier.

Signals received from the broadcast station are processed by the radio tuner. This processing includes selection of one specific broadcast frequency, amplification of the received signal, and removal of the modulation information. The modulated information is in the audio-frequency range. It is presented to the input of the audio amplifier for processing. Eventually, it is reproduced as sound waves in the speakers.

Microphones. The microphone is a device that changes sound waves into electrical waves. There are many different microphones. These include the dynamic, ceramic, and condenser types. These names refer to the type of element that is used to convert the sound waves. Each uses a different electronic principle in order to create the electrical signal voltage.

Dynamic microphones. This microphone uses the principle of electromagnetic induction to generate a signal voltage. The mechanical parts of the dynamic microphone are shown in Figure 5-6. This shows only a part of the microphone housing. Dynamic microphones are available in several different case styles. The microphone consists of a diaphragm, a voice coil, and a

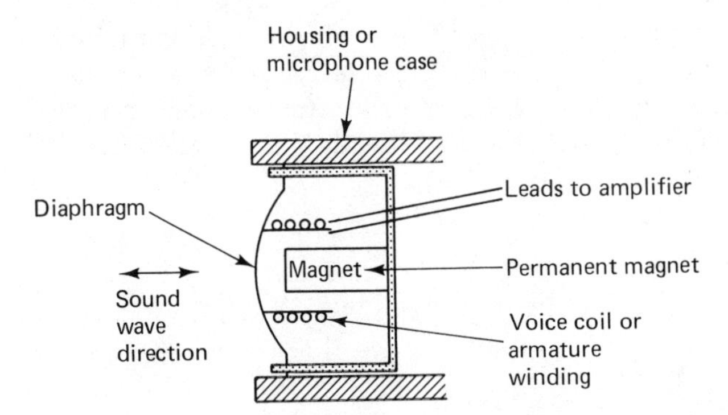

Figure 5-6 Dynamic or electromagnetic microphone.

permanent magnet. The diaphragm is a very thin and flexible piece of metal. It is held in place by the case of the microphone. Sound waves will make the diaphragm flex. A voice coil is fastened to the back side of the diaphragm. This voice coil is made of several turns of fine wire. The wire is wound on a form. The form is fastened to the back side of the diaphragm. A permanent magnet is fastened to a back plate in the microphone. It is made to fit inside the round voice-coil form.

Sound waves that hit the diaphragm cause it to move. This makes the voice coil move back and forth around the permanent magnet. The magnetic field that surrounds the magnet produces a voltage in the voice coil as it moves. This voltage is the output signal of the microphone. Its shape is determined by the waves that make the diaphragm move. Its amplitude is determined by how much voice-coil movement occurs.

Ceramic microphones. A second type of microphone has a ceramic element. A drawing of the parts for this microphone is shown in Figure 5-7. The ceramic microphone is a refinement of the crystal-element microphone. Both operate on the same principle. This is the principle that when a crystal material is distorted, it will produce a voltage. It is called the *piezoelectric effect.*

The diaphragm is held in place by the microphone housing. A push rod is attached to the inside of the diaphragm. The push rod is also connected to the end of a piece of ceramic microphone element material. The other end

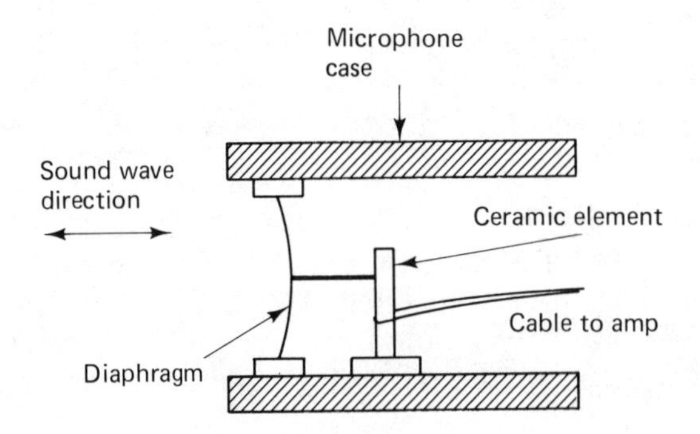

Figure 5-7 Ceramic microphone.

of the ceramic element is held firmly in place in the microphone case. Sound waves push or pull against the diaphragm. This action distorts or changes the shape of the ceramic element. The distortion produces a voltage in the element. This voltage is carried by wires to the input of the audio amplifier.

Condenser microphones. A third type of microphone is called the condenser microphone. This type of microphone is a fairly recent development. It uses the principle of capacitance between two charged plates to create an electronic signal. A drawing of this unit is shown in Figure 5-8. The internal portion of the microphone is much simpler than either the dynamic or the ceramic styles. It consists of a diaphragm and a charged back plate. These two metal surfaces form the plates of a capacitor. The movement of the diaphragm changes the amount of capacitance that is created by this arrangement. The capacitor is used as a portion of a resonant circuit. Any changes in capacitance will change the frequency of resonance.

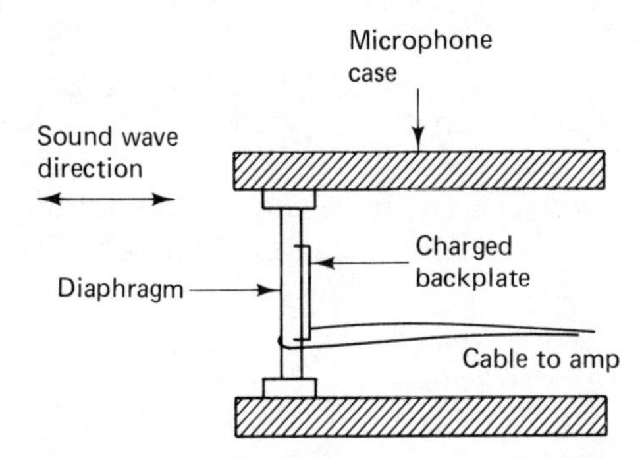

Figure 5-8 Condenser microphone.

A second portion of the condenser microphone system is shown in block diagram form in Figure 5-9. The output of the microphone is connected to an RF oscillator. This oscillator has a frequency of about 10 kHz. The capacitance changes developed by the elements will change this fre-

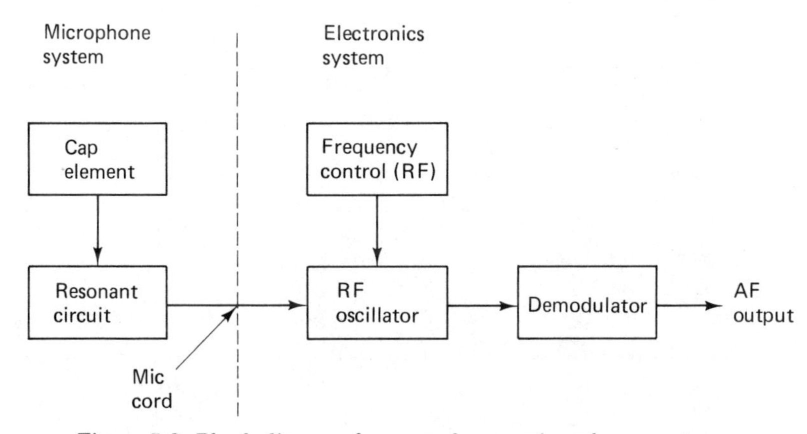

Figure 5-9 Block diagram for a condenser microphone system.

quency. These changes develop as a modulated signal. The audio information is the modulation. A demodulator is used to extract the audio information and present it to the amplifier system. Some condenser microphones have both the element and the electronics in the microphone case. Others separate the two and place the electronics in the basic unit. Many portable tape recorders have both element and electronics built into the tape recorder case.

Phono cartridges. The final major input transducer is the phono cartridge. This device translates physical movements developed in the grooves of a record into an electrical signal. Most phonograph records used today are recorded as stereo. The discussion in this section will use stereo records. Monaural records operate in a similar manner, but have only one channel of information. Figure 5-10 shows the relation of the *stylus*, or needle, to the groove of the record. The grooves on the record are developed to very precise measurements. Each side has an angle of 45° from the vertical plane. The contact area for the stylus is at the sides of the groove rather than at the bottom. The width of the groove depends on the signal impressed at that point. The range of this width is between 0.001 and 0.0026 in. A stylus is developed that will ride in the grooves. The stylus is shown in Figure 5-11. It consists of an arm or cantilever upon which a diamond or other hard-material tip is glued. The cantilever arm is used to transfer the mechanical movement of the stylus in the groove of the record to the transducer device, called the *cartridge*.

The sides of the grooves in the record contain the signal information. The design of the groove places each surface 90° away from each other. This is illustrated in Figure 5-12. Movement that occurs at right angles to the right groove will produce a signal in channel two. Any movement that occurs at right angles to the left-hand groove will produce a signal in channel one. This allows two-channel transcription and playback.

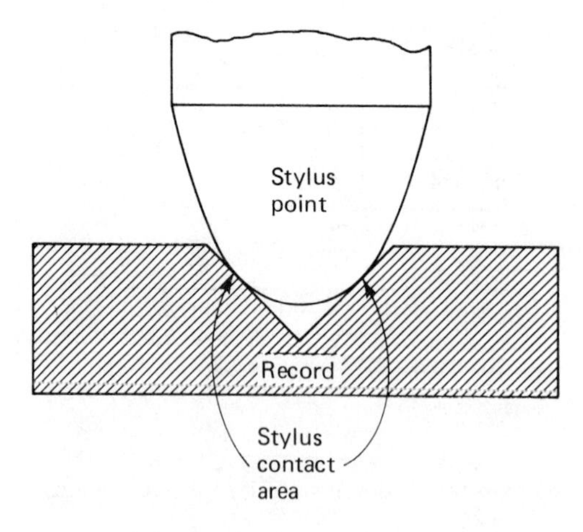

Figure 5-10 The stylus, or needle, must fit correctly into the groove of the record.

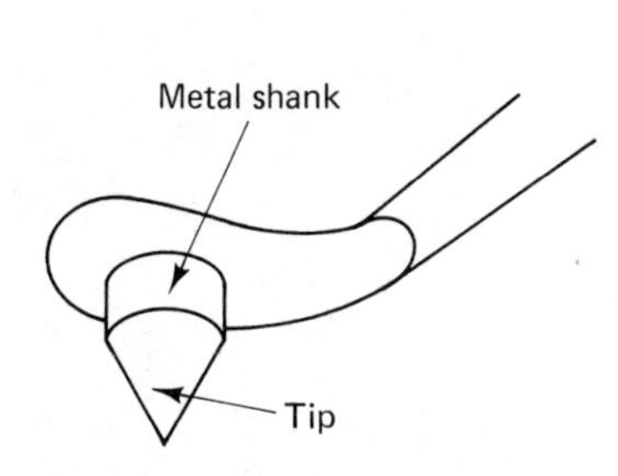

Figure 5-11 Major parts of a phono stylus.

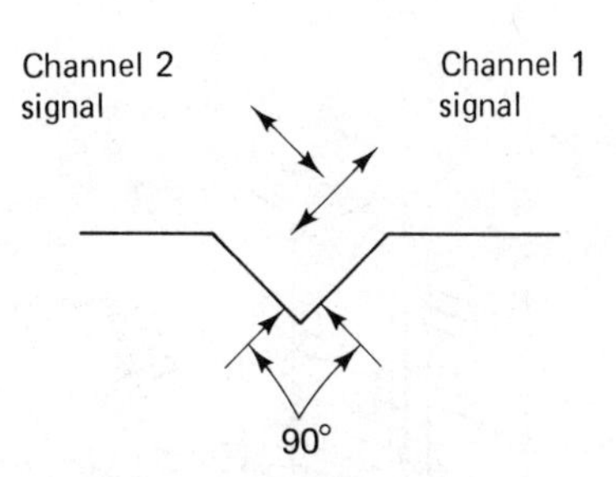

Figure 5-12 Movement of the stylus related to the information recorded in the grooves of the record.

The cartridge is designed so that it will respond to the movements described in the preceding paragraph. Cartridges are made with either ceramic or magnetic transducers. A drawing of the ceramic style of cartridge is shown in Figure 5-13. Two ceramic elements are housed in one container. The elements are arranged so that their axis are about 90° apart. This is necessary because of the way in which the records are manufactured. Movement created by the spacing of the grooves in the record is translated into motion in the cartridge. This motion distorts the ceramic element and produces a voltage. This is the same effect that is developed in the ceramic microphone.

A drawing of a magnetic cartridge element is shown in Figure 5-14. These cartridges are manufactured with either a moving magnet or a moving coil. The electrical principles identified for generators are used here. As long as a magnetic field moves past a wire or a wire moves past a magnetic field a voltage is created. This voltage is similar to that used to create the grooves in the record. Reproduction of the recorded information occurs in this

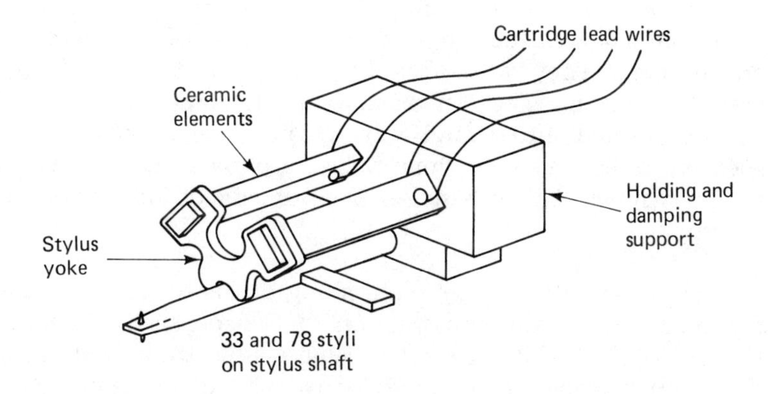

Figure 5-13 Physical layout of a ceramic phono cartridge.

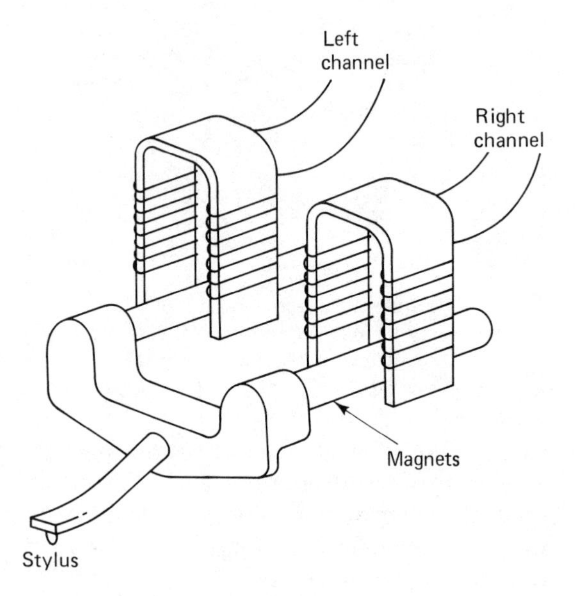

Figure 5-14 Physical layout of a magnetic cartridge.

manner. The signal voltage that is created is then carried to the input to the audio amplifiers.

The devices described in this section are typical of those called *input transducers.* Each uses the basic principles of generating a voltage to convert either sound waves, motion, or magnetic fields into an electronic signal. The signal waveform is similar to that used to create it. The signal requires further processing by an audio amplifier before it can be translated into sound waves. The creation of sound waves is discussed in the next section.

OUTPUT DEVICES

The purpose of the audio-amplifier system is to create sound waves. These waves should faithfully reproduce the form of the signal that is used to create them. They will often be louder due to the efforts of the audio-amplifier system. The electrical waves developed in the amplifier have to be transformed into sound waves. This is accomplished by using a loudspeaker. It can also be done with earphones. Both devices operate in a similar manner. This is based on the principle of interaction of two magnetic fields.

Speakers. The most common output transducer is the speaker. A cross-sectional view of the construction of the speaker is shown in Figure 5-15. It consists of several parts. The frame is used to hold all the components in their proper places. The back (narrow) part of the frame contains a magnet assembly. This is a permanent magnet. It uses iron-pole pieces in order to

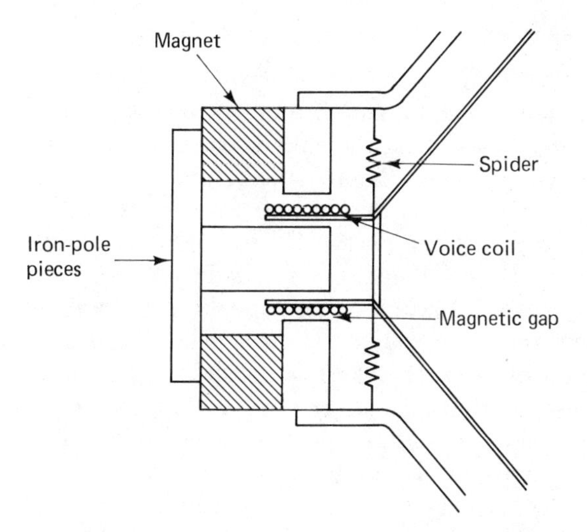

Figure 5-15 Cross-sectional view of a speaker.

concentrate the magnet field in a cylinder at the center of the assembly. A narrow round opening is left at the front end of the core assembly. This opening accommodates the voice-coil assembly.

The *voice-coil* assembly consists of a coil of wire wound on a round form. The voice-coil assembly has a specific impedance and power-handling capability. The impedance of the voice coil is the same as that of the speaker assembly. Common impedances used for speakers are 4 and 8 Ω. Other values are also used, but these two are the most common ones. The power-handling rating is in units of the watt. This describes the maximum amount of audio power the speaker is capable of handling. Operating the speaker with power in excess of this rating will damage the voice coil and thus make the speaker inoperative.

The part of the speaker that actually moves the air in order to create sound waves is called the *cone*. The cone is a conical or funnel-shaped piece of paper. It is designed so that it has a rigid, fixed shape. The smaller end of the cone is attached to the voice coil. The large end of the cone is connected to the frame of the speaker. Some sort of flexible material is used between the large cone end and the frame so that the cone is able to move back and forth in the frame.

The movement of the cone is accomplished by creating a magnetic field in the voice coil. The voice coil is connected by wires to the output terminals of the audio amplifier. The output signal has magnetic properties. It is in the form of a low voltage and reasonably high current. The current flow through the voice coil makes it an electromagnet. The voice coil magnet interacts with the field of the magnet assembly at the back of the speaker. This interaction forces the cone to move. As it moves, it pushes and pulls air. The rate of movement of the air determines the frequency of the sound waves. The distance the cone is moved determines the loudness of the sound waves. A flexible device called a *spider* is used to hold the voice coil assembly in its proper place.

Frequency response. As a general rule the physical size of the speaker determines its frequency response. Large speakers usually are used to reproduce the lower audio frequencies. These speakers are called *woofers.* The high frequency for this speaker is about 800 Hz. A speaker that will efficiently reproduce sounds in the range 800 Hz to 5 kHz is called a *midrange* speaker. The frequencies that are above 5 kHz are reproduced by speakers that are called *tweeters.* Each of these speakers is designed to be most efficient in its own frequency range. Some speakers are capable of reproducing a greater range of frequencies. The exact range is determined by the shape and firmness of the speaker cone. Some systems have two speakers instead of three. The low-range speaker has a higher-frequency response and the high-range speaker is able to reproduce sounds in the midfrequency range. This is accomplished by treating the cone so that it is slightly more flexible, and thus responsive to these frequencies. Another style of speaker has a small cone inside the large cone. This is called a *whizzer* cone. It is capable of reproducing the higher frequencies. Speakers using a whizzer cone are capable of reproducing sounds in the entire audio range.

Crossover networks. When two or three speakers are used in a system a method of directing signals to the proper speaker is required. This is accomplished with a device called a *crossover network.* The most common crossover networks are filter systems. A schematic for one is shown in Figure 5-16. It consists of three filter networks. Each filter is a bandpass filter that is designed to allow only specific frequencies to pass from the output of the amplifier to the speaker. Crossover networks can be designed

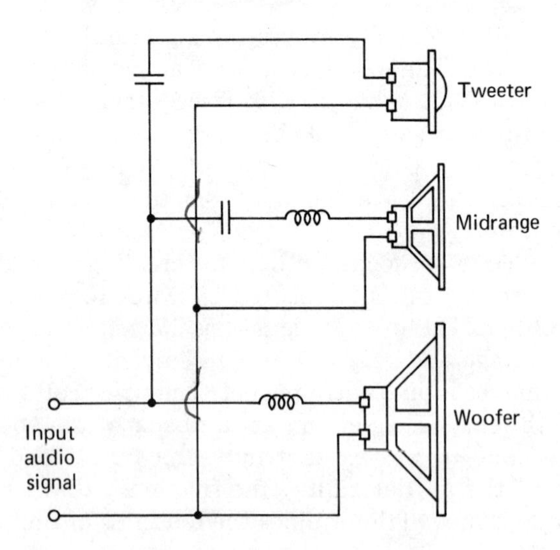

Figure 5-16 Schematic diagram of a three-speaker system with crossover networks.

to pass any group of frequencies desired. The specific range is left to the designer of the system.

Impedance matching. The efficiency of any system is determined by its ability to minimize losses due to undesired heat or friction. This is true in all electronic systems. Maximum power should be transferred from the amplifier to the speaker system. Losses due to heat must be held to a minimum. This can only be accomplished when the impedance of the speakers matches that of the output of the amplifier.

All amplifiers are rated with an output impedance. This is usually 4, 8, or 16 Ω. Some amplifiers have all three output impedances available. Speakers usually are rated at either 4 or 8 Ω impedance. When a single speaker is connected to the amplifier, selection of the proper impedance is easy. When more than one speaker is connected to the amplifier, the system is a little more complicated. The rules for resistances in series and parallel apply to speaker wiring. This is shown in Figure 5-17. The four 8-Ω speakers connected in series as shown in part (a) have a total impedance of 32 Ω. This is a serious mismatch for an 8-Ω amplifier output connection. The result is a large heat loss and low audio output. When the same four speakers are parallel connected as shown in part (b) of this figure, the effective impedance is 2 Ω. This is also a mismatch. The low impedance permits large current flow through the output of the amplifier. It may damage the output transistors or, at the very least, blow a protective fuse.

The best way to connect the four speakers is shown in part (c) of this figure. A combination of series and parallel wiring is used. The result is an 8-Ω impedance. This matches the output impedance of the amplifier and losses due to mismatch are minimized. Maximum power is transferred from the amplifier to the speakers in this manner.

Headphones. Headphones are miniature speakers. They are also personal because under proper operating conditions only the wearer can hear the sounds they produce. The construction of a headphone is shown in Figure 5-18. This is a dynamic, or electromagnetic, type of reproducer. The signal from the amplifier is connected to the field coil windings. These are mounted into the frame of the headphone. A thin metallic diaphragm is also attached to the frame. It is arranged so that there is a small air gap between it and the pole pieces of the electromagnet field coils. A current from the amplifier passes through the field coils during operation. This makes them electromagnets and they pull the diaphragm toward the pole pieces. Alternate pulling and releasing of the diaphragm establishes sound waves. This effect reproduces the sound signal.

Headphones may be connected in series or parallel or they may be separately wired to the amplifier. This is illustrated in Figure 5-19. Part (a) of the drawing shows the monaural connection. Either series (upper) or

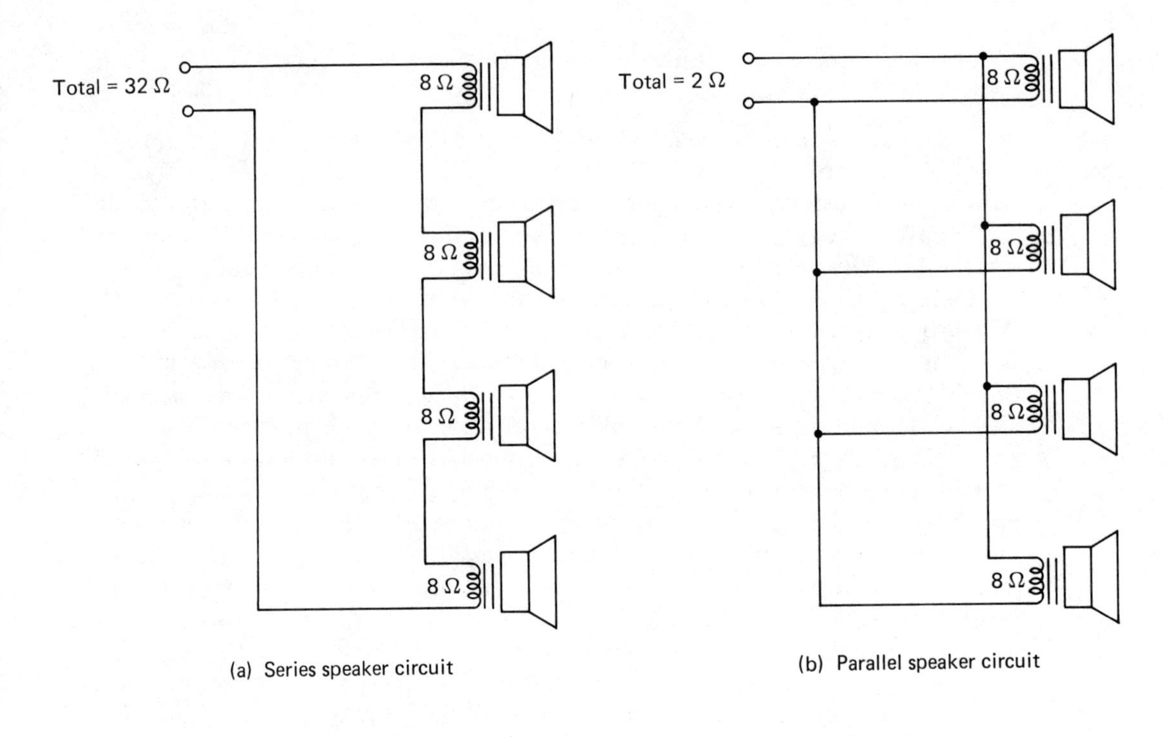

(a) Series speaker circuit

(b) Parallel speaker circuit

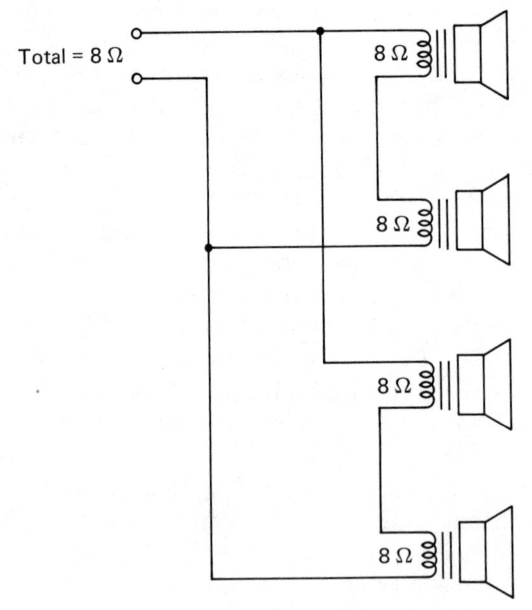

(c) Combined series and parallel speaker circuits

Figure 5-17 Multiple connections for speakers in order to match amplifier output impedances.

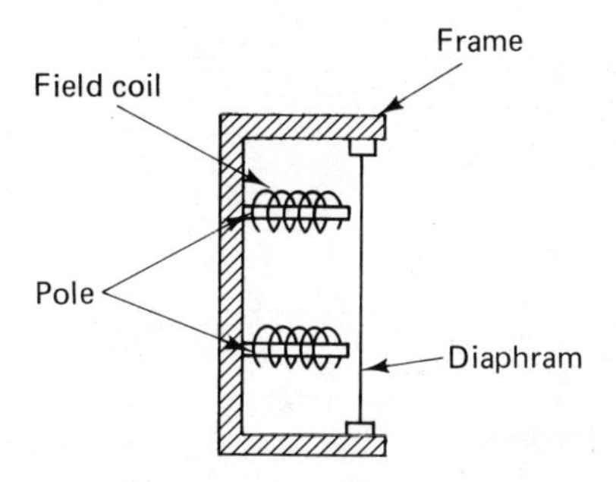

Figure 5-18 Cross-sectional view of the construction of a headphone.

parallel (lower) connection may be utilized. The specific wiring arrangement depends on the requirements of the system. A stereo connection is shown in part (b) of the illustration. This system has one earphone connected to the output of one channel and the second earphone connected to the output of the other channel. Often, three wires are used instead of four in this system. One lead from each earphone is connected to circuit common when this system is used.

The wiring of connecting plugs for stereo and monaural headphones is shown in Figure 5-20. The upper diagram shows a stereo connection. The plug has three circuits. These are identified as the tip, the ring, and the sleeve. Each is insulated from the other by a piece of plastic. The wiring connections are all terminated in the rear of the plug assembly. Some of these are solder connectors, whereas others utilize a screw to hold the wires in place. The tip and the ring are each connected to one of the field coil windings. The sleeve is connected to the common wire for the two field coils. The monaural plug has two circuits instead of three. One of these is

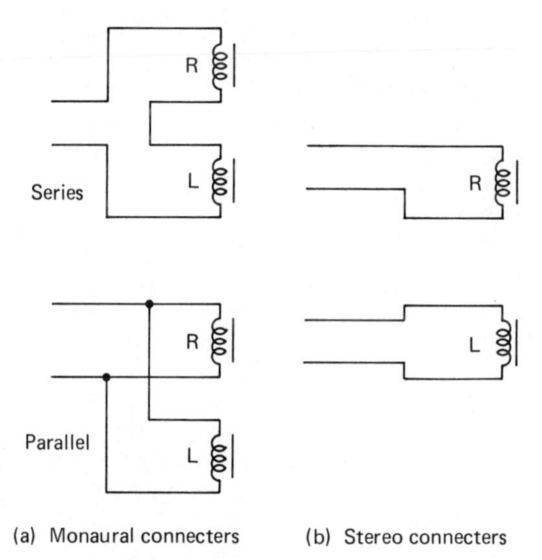

Figure 5-19 Headphone wiring: (a) monaural circuits show both series and parallel connections; (b) stereo circuits require three wires.

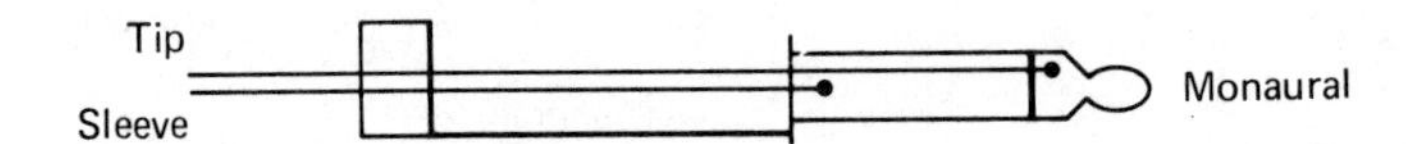

Figure 5-20 Connecting plug wiring for stereo and monaural headphone plug.

the tip and the other is the sleeve. Here, too, the sleeve is connected to circuit common and the tip is the signal lead.

This completes the discussion about input and output devices. Each stereo system must use at least one of the input transducers. Each stereo system utilizes one of the output transducers unless the signal is to be stored on a record or a tape. An understanding of how these devices are utilized to produce a signal voltage is essential for the successful electronic service technician.

QUESTIONS

5-1. Name four kinds of amplifier input devices.
5-2. How is the signal transferred from the tape to the head?
5-3. How is the signal transferred from the airwaves to the antenna?
5-4. How is the signal transferred from the record to the phono cartridge?
5-5. How is the signal transferred from the output stage to the speaker cone?
5-6. What frequencies are reproduced by a woofer, a midrange, and a tweeter speaker?
5-7. Why is impedance matching important in a speaker system?
5-8. What is a speaker crossover network?
5-9. What is the basic electronic principle used by a condenser microphone?
5-10. What is the basic electronic principle used by a dynamic microphone?

Chapter 6

AM and FM Tuners

One of the major input devices used in stereo systems is the radio-frequency tuner. This device is better known as either an AM or an FM tuner. In a good number of stereo units these tuners are built into the same cabinet as the audio amplifier. When this is done the units are called *integrated receivers*, or just *receivers*. The material presented in this chapter covers the operation and function of AM, FM, and FM stereo tuners.

GENERAL REQUIREMENTS

A tuner's purpose is to select, amplify, and demodulate the broadcast station signal. Several factors are involved in this process. Each is important for the successful operation of the tuner.

Selection. The air around us is full of radio-frequency energy. Broadcast stations operate on many frequencies above the high end of the audio-frequency spectrum. These broadcast stations provide an assortment of plex multichannel telemetry signals used for space satellites. The specific frequencies and modes of operation are determined and controlled by both national and international agreement. In the United States all broadcasting is regulated by the Federal Communications Commission (FCC) in Washington, D.C.

Certain groups of frequencies are assigned for specific purposes. One of these is called the AM radio band. Its frequency spectrum starts at 540 kHz and increases to 1600 kHz (1.6 mHz). AM broadcast stations are assigned to *channels* or frequencies that center on each 10 kHz of the band. This means that the tuner should be able to select one station and reject any other stations operating near the desired frequency.

The ability of a tuner to pick out one signal from the many being broadcast is called its *selectivity*. Since each AM broadcast station is separated from an adjacent station by 10 kHz, the width of the received signal should not be greater than 10 kHz, or 5 kHz on either side of the assigned frequency of the broadcast station. This is accomplished by using resonant circuits in the tuner. These circuits have a bandwidth of 10 kHz in order to do this.

Figure 6-1 illustrates the bandwidth of a modulated RF signal. The carrier has a very narrow bandwidth. It does not extend beyond its assigned frequency. The modulation process increases the bandwidth of the broadcast signal. Frequencies used for modulation are both added and subtracted from the carrier signal. This causes an increase in the bandwidth of the broadcast signal. It increases by the amount of the frequency of the modulation. AM broadcast stations are limited to a bandwidth of 10 kHz. This means that the modulating signal frequency cannot exceed 5 kHz. The 5 kHz above and 5 kHz below the carrier frequency occupy a space that is 10 kHz in width. If a tuner did not limit its bandwidth to 10 kHz for an AM signal, it would select and reproduce the signal from more than one broadcast station at the same time. This situation would be disturbing to the listener, to say the least!

The discussion so far has dealt with the AM tuner bandwidth. Different conditions exist for the FM tuner. The modulated signal for an FM station

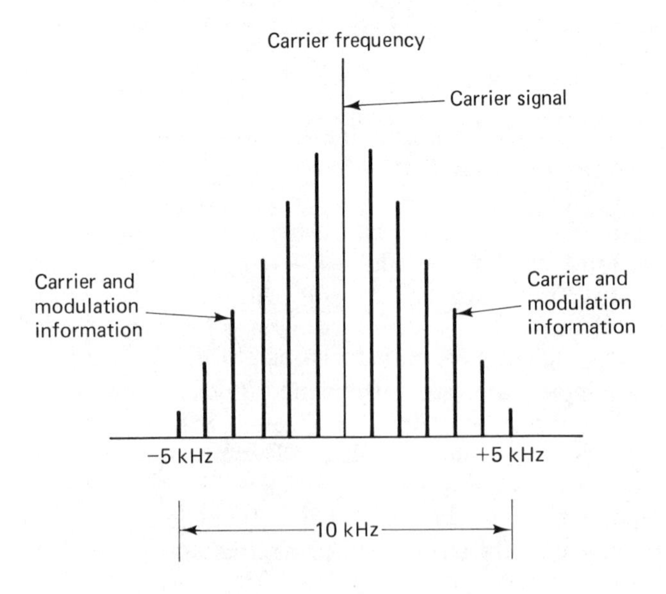

Figure 6-1 Spectrum analysis of the bandwidth of a modulated RF signal.

requires up to 150 kHz of space. This breaks down to be 75 kHz above and 75 kHz below the frequency of the broadcast station carrier frequency assignment. The selectivity for an FM radio has to be 150 kHz in order to process the broadcast information successfully.

Sensitivity. The ability of a tuner to pick up a broadcast signal is called its *sensitivity*. The broadcast signal has electromagnetic properties. This signal crosses radio or tuner antenna wires. When the broadcast signal crosses the antenna wires it induces a voltage on those wires. This voltage is often in the microvolt (μV) range. Design determines the sensitivity of the tuner. The purpose is to achieve a level of signal output from the tuner that is adequate for quality reproduction by the audio amplifier. The term normally used is *full quieting*. This indicates a high level of signal and a very low level of background noise. It is called the *signal-to-noise ratio* of the receiver.

An AM tuner's sensitivity will range from 100 to 1000 μV. The exact value is determined by the design of the tuner. Some AM–FM stereo tuners have poor-quality AM sections. This is evidently a deliberate design situation. Evidently, some radio designers feel that the limited frequency response of the AM broadcast signal and the rapid acceptance of FM stereo broadcasting has limited the need for a good-quality AM tuner.

The FM tuner is much more demanding. Its sensitivity requirements are on the order of 20 μV or less. The FM tuner has to have more stages of amplification than does the AM tuner in order to accomplish this. Let us look at the requirements of both the AM and the FM tuner in order to understand how the sections of these devices function.

AM TUNER

A block diagram of an AM tuner is shown in Figure 6-2. This type of tuner is called a *heterodyne* tuner. This is because of the heterodyning, or mixing action, that occurs. The mixing action is discussed in the next paragraphs.

RF amplifier. The purpose of this block is to select and amplify the desired RF signal. This block contains a tunable resonant circuit. It receives the signal from the antenna and amplifies the selected frequency. Many economy AM tuners do not have an RF amplifier block. The tuned circuit feeds the signal directly to the next block without providing any amplification.

Local oscillator. The purpose of this block is to create a carrier wave (CW) signal. This block also contains a tunable resonant circuit. The design of the local oscillator is that its frequency differs from that of the received signal. The amount of this difference for an AM tuner is 455 kHz. The RF

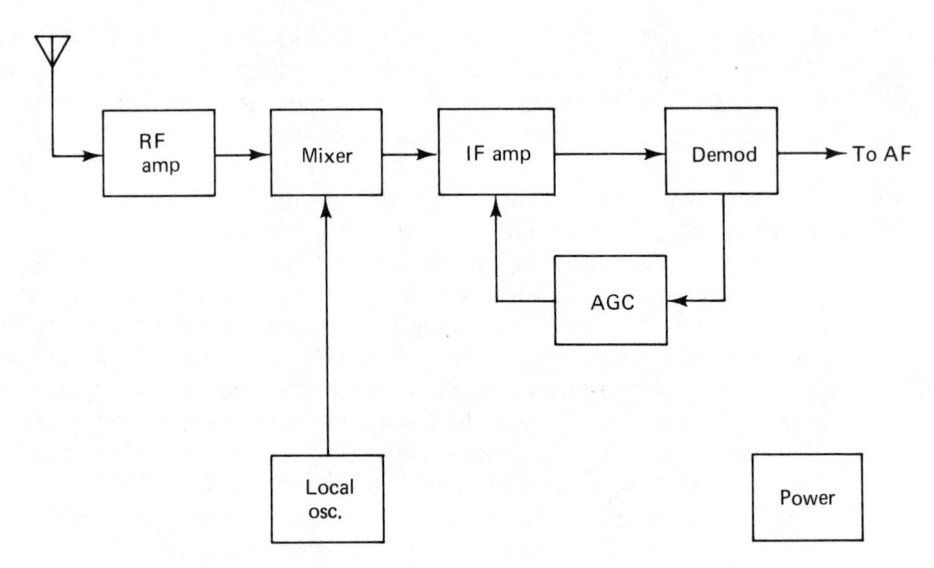

Figure 6-2 Block diagram for an AM radio tuner system.

amplifier and the local oscillator tune together. This *tracking* of the two sections will always produce a difference in frequency of 455 kHz. An exception to this is the 262-kHz IF frequency used in many car radios.

Mixer. Mixing action, or heterodyning, occurs in this block. The mixer has two inputs and one output. The two inputs are the selected RF frequency and the local oscillator CW signal. Mixing action actually produces several frequencies. There are four major signals present at the output of the mixer. These are shown in Figure 6-3. These are both of the original signals, a signal whose frequency is the sum of the two original signals and a signal whose frequency is the difference between the two original signals. When one of the input signals is modulated, the sum and difference output signals also contain the modulation. The selection of the desired output signal is a design factor. Both the sum and the difference signals are utilized in various tuner circuits. The important fact for the AM and FM receiver is that a modulated signal with a frequency that is the mathematical difference between the two input signals is developed.

IF amplifier. The third block in this system is called the *intermediate-frequency* (IF) amplifier. This block has several fixed-tuned resonant circuits. They are tuned to 455 kHz in the AM tuner. Much of the amplification of the signal being processed occurs in this block. This is deliberately done to avoid the need for the consumer to do this tuning. The bandwidth for the 455 kHz IF amplifier is 10 kHz. It is called the IF because it has a frequency that differs from either the one being received or the one created by the local oscillator in the receiver. This is another or intermediate, frequency.

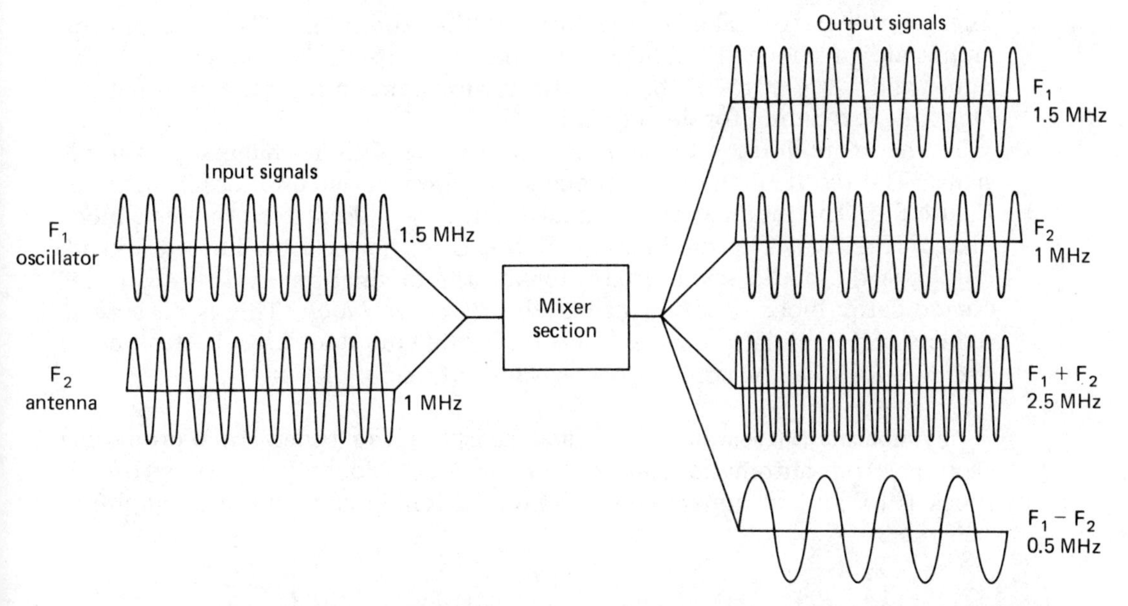

Figure 6-3 Signals available at the output of the mixer stage. (From Joel Goldberg, *Radio, Television, and Sound System Repair: An Introduction,* © 1978, page 123. Reprinted by permission of Prentice-Hall, Inc.)

Demodulator. The next block in the signal-processing path is the demodulator. The function of this block is to extract the modulating information from the composite carrier and modulation signal. The waveform of an amplitude-modulated carrier wave is shown in Figure 6-4. A carrier and an audio signal are required in this example. The modulation process in the transmitter changes the carrier signal as shown. The upper and lower out-

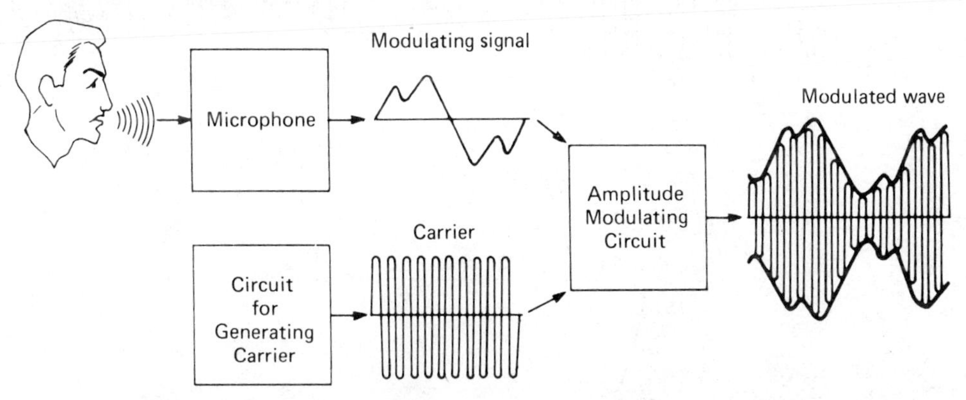

Figure 6-4 An amplitude-modulated carrier assumes the shape of the modulating signal. (From Joel Goldberg, *Radio, Television, and Sound System Repair: An Introduction,* © 1978, page 90. Reprinted by permission of Prentice-Hall, Inc.)

lines of the carrier take on the form of the modulation. This changes the amplitude of the carrier, hence the name "amplitude modulation." The upper and lower edges of the modulated wave contain the same information. Only one is required for demodulation.

The demodulator block reacts to the amplitude changes of the IF signal. The result of the signal-processing action of an AM tuner is shown in Figure 6-5. The modulated RF signal and the local oscillator produce a modulated IF signal. The modulation of the IF signal is demodulated and its output is the audio signal used to create the original signal. The AM tuner demodulator block is often called the *detector block*. This is because it detects the modulation signal. The term "demodulator" is a better one to use because it can be applied to all systems of demodulation.

Automatic gain control. The final block in the tuner signal-processing section is the automatic gain control, or AGC, block. The purpose of this block is to control the amount of amplification, or gain, in the IF amplifier.

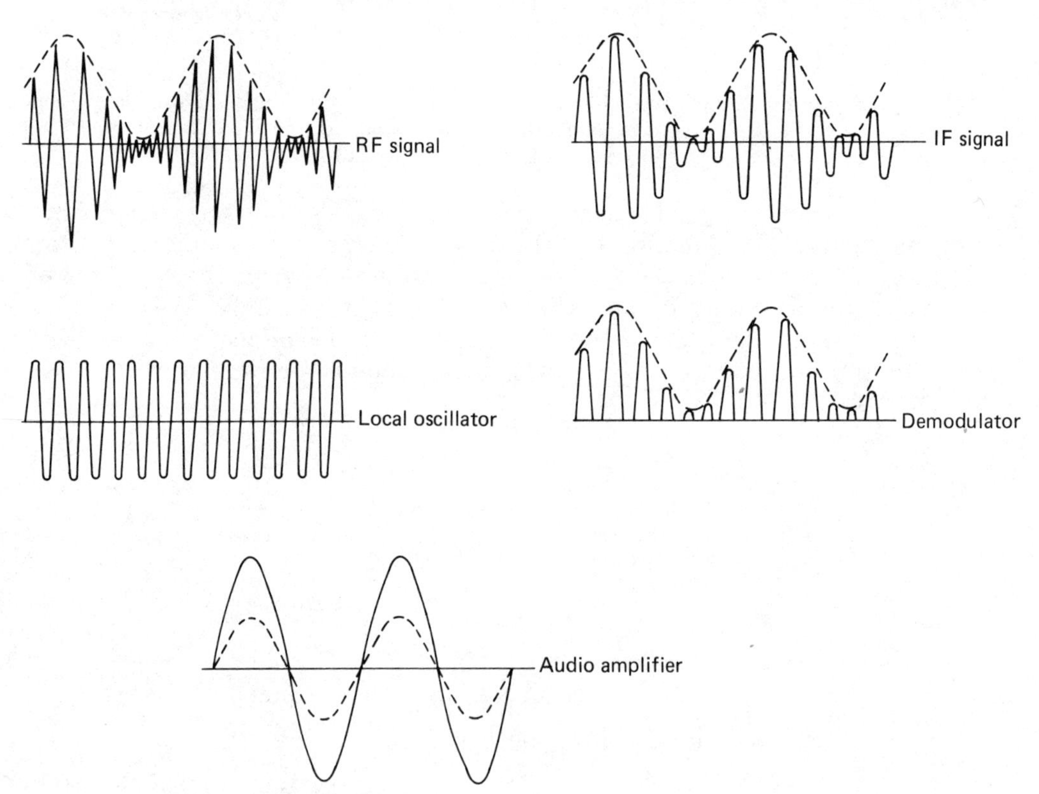

Figure 6-5 Demodulator action of the AM signal. (From Joel Goldberg, *Radio, Television, and Sound System Repair: An Introduction*, © 1978, page 125. Reprinted by permission of Prentice-Hall, Inc.)

This is necessary so that the signal at the demodulator is great enough to permit a high signal-to-noise ratio. A portion of the demodulated signal is processed in the AGC block. The output of the AGC block is a dc control voltage. This voltage is used to vary the operating bias on the IF amplifier. The gain of the IF amplifier is adjusted constantly by the AGC output.

Power source. All electronic devices require operating power. The purpose of this block is to provide the proper voltages and currents for successful tuner operation. The source may be batteries, ac, or even sunlight or solar panels.

FM TUNER

The block diagram for an FM stereo tuner is shown in Figure 6-6. Several of the blocks used in the AM tuner are also used in the FM stereo tuner. These blocks also have very similar functions. Only those blocks that have significant differences are described in this section.

RF amplifier. This block's major difference is in the frequencies of operation. The FM radio band occupies a frequency spectrum of 20 MHz. Its assigned frequency ranges from 88 to 108 MHz. The bandwidth of an FM stereo signal is 150 kHz. Each broadcast station occupies a space that extends 75 kHz on either side of its assigned carrier frequency. The reason for this is that the FM stereo signal is more complex. In order to broadcast all the stereo information, the wider bandwidth is required.

Local oscillator. The local oscillator creates a CW signal. The frequency of this signal is between 98.7 and 118.7 MHz. The oscillator and the RF amplifier are tuned together, as in the AM tuner.

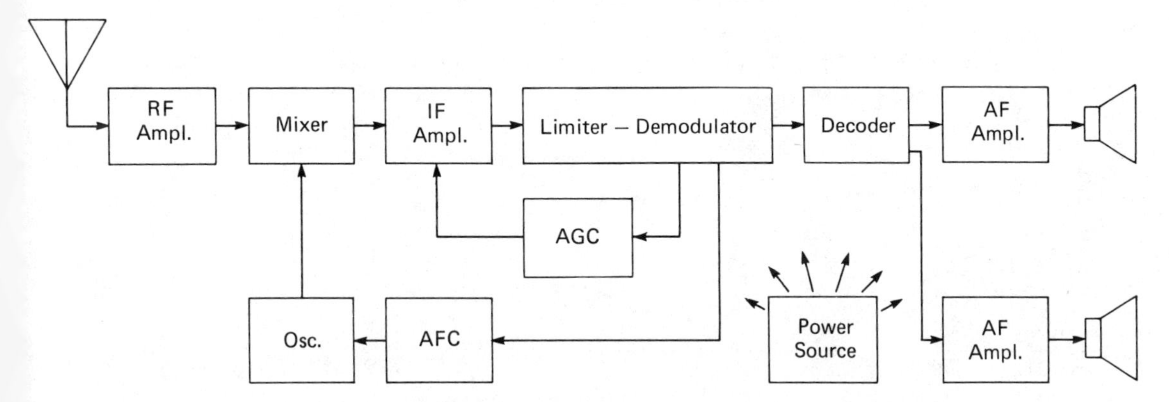

Figure 6-6 Block diagram for an FM stereo tuner system.

Mixer. The output from the mixer is a frequency-modulated signal. The carrier frequency for this signal is 10.7 MHz. This is a constant value. It is produced when the RF and oscillator signals are heterodyned in this block.

IF amplifier. This block provides a great amount of amplification for the modulated 10.7-MHz IF signal. It is fixed tuned and has a bandwidth of 150 kHz in order to pass all the desired signal.

Limiter. This block is one that is not used in the AM tuner. Its function is to limit the amplitude of the FM IF signal. Amplitude variations of the signal, caused by electrical discharges in the air or other factors, are not desirable in the FM tuner. The limiter block cuts off, or limits, the amplitude of the FM signal to a level that stops these AM signals from being processed. Many tuners do not use a separate limiter block. Limiting action is done in the demodulator block instead.

Demodulator. This block's purpose is to remove the modulation from the IF signal. The modulation signal of an FM stereo transmitter contains much more information than that found in the AM signal. An analysis of the FM stereo signal is shown in Figure 6-7. The first 15 kHz on either side of the carrier contains a frequency-modulated monaural signal. This signal is called the *L + R signal*. It is the signal that is processed by FM monaural receivers. A small tone is impressed on the carrier at the 19-kHz point. This tone is called the *pilot signal*. It is used as a reference for the FM stereo carrier signal.

The FM stereo signal is a multiplex signal. This means that it has a carrier frequency of its own as well as modulation. The FM stereo component is an AM signal. It uses a suppressed carrier modulation system and has both an upper and a lower sideband. It has a bandwidth of 30 kHz. The frequency of the suppressed carrier is 38 kHz. This signal is added to the station carrier by a process called *multiplexing*. This process is used to add additional signals on to the station carrier signal. These signals are added at frequencies that do not interfere with the basic modulation system. A fre-

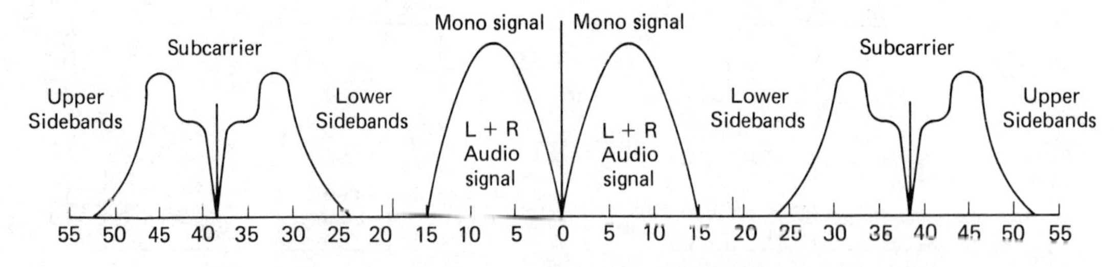

Figure 6-7 Spectrum analysis of the composite FM stereo signal.

quency of 38 kHz is well above the audio range. These signals are carried to the receiver and are not processed for monaural reproduction. An additional demodulator is required to make them usable in the stereo tuner. The stereo component is called the *L - R signal.*

The purpose of the 19-kHz signal is that its second harmonic is used to recreate the suppressed 38-kHz carrier for the stereo L - R component. Demodulation of a suppressed carrier signal cannot occur unless the carrier is recreated as a part of the demodulation process. The reason for the use of a 19-kHz signal is that it is "clean". In other words, there are no other signals near it on the carrier. It is not possible to have any interference with this signal as it is processed in the tuner.

The FM demodulator does not react to the stereo multiplex AM signal. The multiplex signal passes through the FM demodulator as a part of the demodulated FM carrier. The information contained in the stereo signal requires further processing before it can be used. This is done in the stereo decoder block described later in this chapter.

Automatic gain control. The AGC block uses a sample of the demodulated monaural signal. It is converted into a dc control voltage. This control voltage is fed back to both the IF amplifier and the RF amplifier. This control voltage is used to adjust the gain of these blocks.

Automatic frequency control. This block is used to maintain the frequency of the local oscillator in the tuner. Automatic frequency control (AFC) is the development of a second dc control voltage. This also is taken from the demodulated monaural signal. It is fed back to the local oscillator frequency-determining components. This process is necessary because these components have a tendency to change in value as their operating temperature changes. Many of these components are selected so that their values will remain constant under thermal changes. There still is a tendency for a frequency change. AFC is used to correct this change and maintain the oscillator frequency. This, in turn, will keep the tuner on its selected frequency.

Power source. This tuner is similar to the operation of other electronic devices in that it requires specific operating voltages and currents. The power source develops these values.

Stereo decoder. The stereo decoder is composed of several blocks. These are shown in Figure 6-8. A composite signal is demodulated by the tuner demodulator block. This signal contains the monaural signal (L + R), a 19-kHz pilot signal, and the suppressed carrier stereo signal (L - R). There are three major occurrences in the decoder. All happen at the same time. Let us examine them one at a time, starting at the demodulator. The composite stereo signal is sent in three directions. One of these is through a low-pass

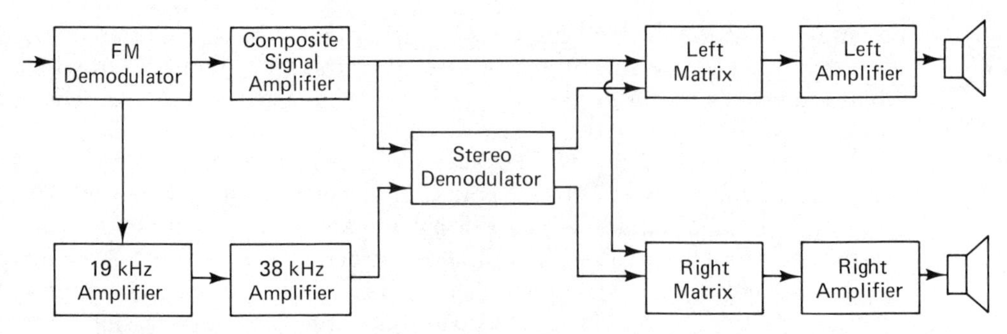

Figure 6-8 Block diagram for one form of FM stereo decoder system. (From Joel Goldberg, *Radio, Television, and Sound System Repair: An Introduction*, © 1978, page 51. Reprinted by permission of Prentice-Hall, Inc.)

filter. This filter has a cutoff frequency of 15 kHz. It will pass only those frequencies that are below this cutoff point. The signal then passes through a delay line. Here it is slowed a few fractions of a second. This is necessary because of the processing time required for the rest of the stereo signal. Once the signal leaves the delay it is sent to two blocks. These are called *adders* or matrixes. In these blocks the monaural signal is combined with the stereo component. The output of these adders is connected to channels of the audio amplifier.

A second path from the demodulator is used for the 19 kHz signal. The composite signal is sent to a 19 kHz amplifier. This amplifier only reacts to a signal frequency of 19 kHz. All others are rejected and stop at this point. The 19 kHz signal is sent to a 19 kHz oscillator. It is used to synchronize the oscillator frequency with that of the broadcast signal. The output of this oscillator is frequency doubled to 38 kHz. The 38 kHz signal is then sent to a stereo signal demodulator. It will be used to demodulate the stereo component.

The third path for the composite stereo signal takes it through a band-pass filter. The frequency range of this filter is from 23 to 53 kHz. This permits the L - R stereo signal to pass and rejects all other signals. The L - R signal is then presented to the stereo demodulator. When both the L - R and the 38-kHz carrier signal are present, the demodulator will operate. It has two outputs. These are the same signal. One is 180° out of phase with the other. The phase reversal is done by a phase inverter in the demodulator. The output signals for this block are a +(L - R) and a -(L - R) signal. Removal of the parentheses will give us a +L - R and a -L + R signal.

The two demodulated signals are next sent to the adder blocks. One signal is sent to each of the blocks. At these blocks they are added to the monaural signal from the delay line. The results are:

$$\begin{array}{l} +L + R \text{ monaural} \\ \underline{+L - R} \text{ stereo} \\ 2L \;\; 0R \text{ left-channel stereo} \end{array}$$

and

$$\begin{array}{l} +L + R \text{ monaural} \\ \underline{-L + R} \text{ stereo} \\ 0L + 2R \text{ right-channel stereo} \end{array}$$

These two signals are then connected to the appropriate audio amplifier. The system may sound complex but it is necessary if one is to receive FM monaural signals with a stereo tuner. The absence of a stereo signal provides monaural information to both audio output circuits.

This completes the processing of the AM and FM stereo signals. The next section covers tuning units. This refers to the RF amplifier, local oscillator, and mixer sections of the tuner.

ELECTROMECHANICAL TUNING SYSTEM

For many years the solution of the desired frequency for reception of electronic signal was accomplished by varying one component in a resonant circuit. The systems used are shown in Figure 6-9. This illustrates a resonant circuit that utilizes a variable capacitor and a fixed-value inductor. Almost all radios and tuners used in the home utilized this type of circuit. A dual-section variable capacitor is used to resonate the circuit over the entire AM or FM broadcast band. The dashed lines on the partial schematic indicate

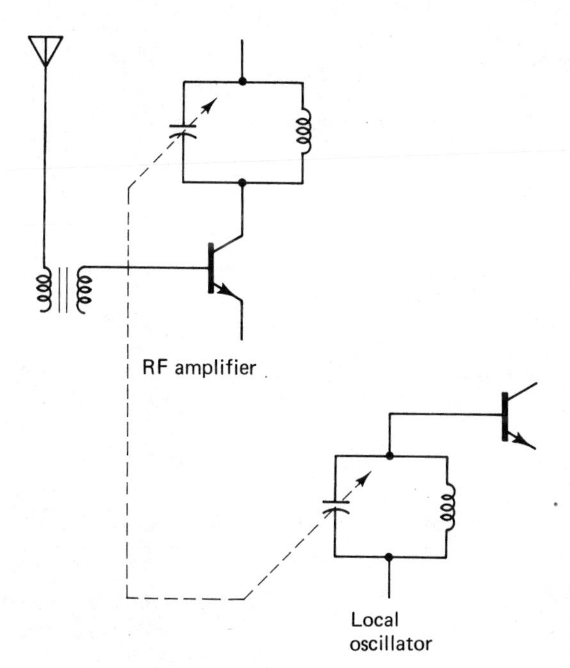

Figure 6-9 Tunable capacitors in resonant circuits commonly found in all heterodyne receivers.

that the two capacitors are electrically separate but physically connected to each other. The rotors of both capacitors are built onto the same shaft. Turning the shaft changes both values at the same time. The electrical characteristics of the local oscillator are such that its circuit resonates at a frequency that is 455 kHz above the RF circuit of the AM radio. Oscillator circuits used for reception of FM oscillate at a frequency that is 10.7 MHz above the desired received signal frequency.

A second method of tuning a resonant circuit uses variable inductors rather than variable capacitors. An example of this type of circuit is shown in Figure 6-10. Two inductors are shown in this illustration. In actual practice these or more are used. The drive shaft is connected to a worm gear. The slugs, or cores, of the inductors are connected to a drive bar. Rotating the tuning shaft makes the drive bar move. This, in turn, moves the cores in and out of the inductors. This, when used with a fixed value of capacitance, will provide a variable resonant circuit.

The main application for the variable inductor circuit is in automobile radios. The reason for this is very simple. Automobiles operate in a dirty environment. Dust and other foreign matter is often found in the air. Dust has a tendency to lodge between the plates of a variable capacitor. It will either change the amount of capacitance, develop a noisy tuning system, or short out the capacitor plates. Any one of these is undesirable. Use of a vari-

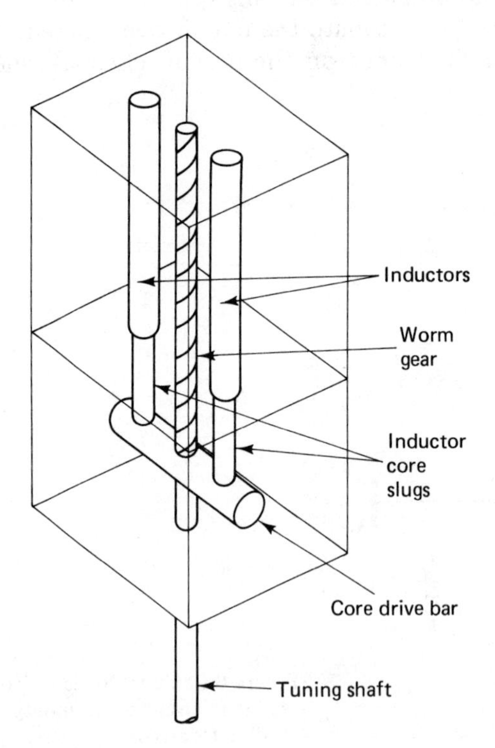

Figure 6-10 Tunable inductors used in heterodyne receivers.

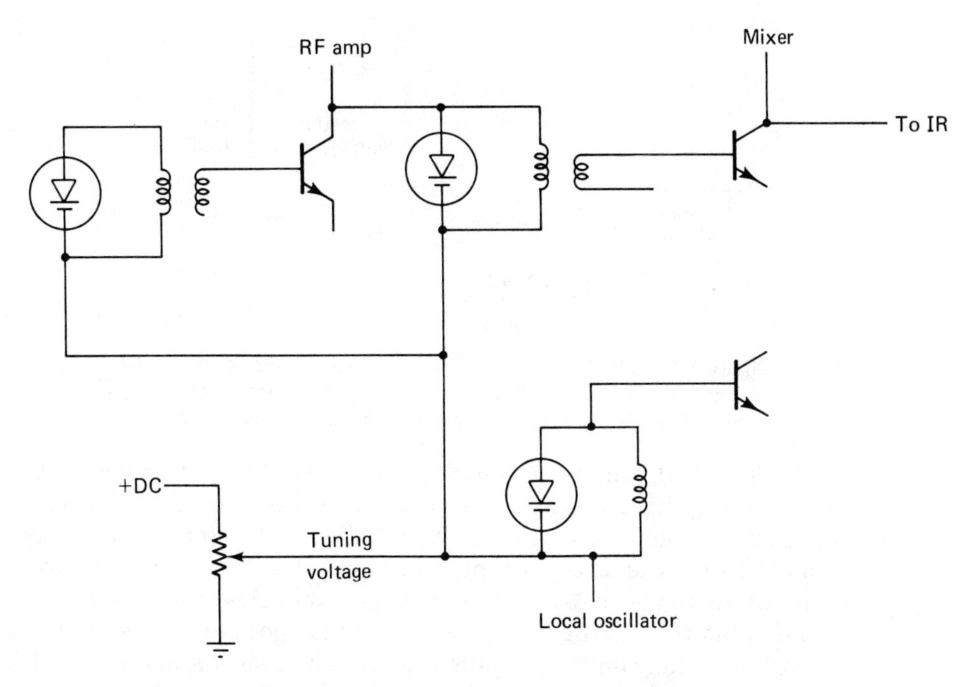

Figure 6-11 Varactor diodes used as voltage-variable capacitors in resonant circuits.

able inductance with a fixed value of capacitance eliminates most of these problems.

A modification of the movable capacitor or inductor method of tuning uses varactor diodes. The advantage of the varactor diode is that it eliminates almost all of the need for mechanical tuning parts. The only component that moves in a varactor diode tuning circuit is a potentiometer. A circuit using varactor diode tuning methods is shown in Figure 6-11. This shows only that part of the circuit using varactor diode tuning. A potentiometer is connected to the dc supply. The arm of the potentiometer is connected to each of the varactor diodes. Changing the level of dc voltage applied to the varactors changes their capacitance value. Since the varactor diode acts as a capacitor a resonant circuit is formed using it and an inductance. Varying the capacitance will vary the resonant point of the circuit. Modifications of this type of circuit are used with all electronic integrated-circuit-controlled tuning.

ELECTRONIC TUNING SYSTEMS

The development of the integrated circuit (IC) and its associated components has led to a completely different type of tuning control circuit. The name used to describe this newer tuning circuit is *phase-locked-loop* (PLL)

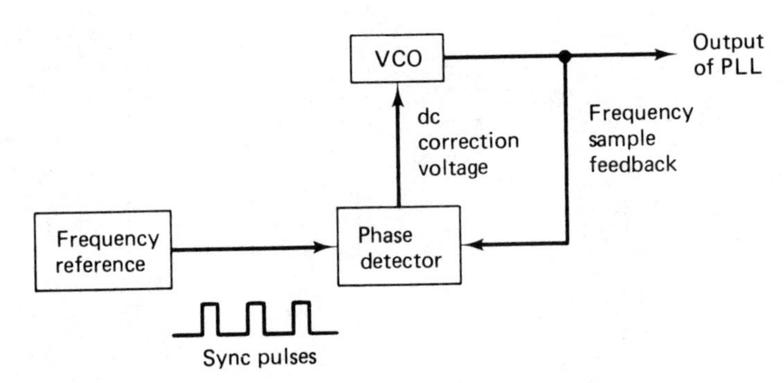

Figure 6-12 Block diagram for a phase-locked-loop (PLL) system. (From Joel Goldberg, *Fundamentals of Television Servicing,* © 1982, page 214. Reprinted by permission of Prentice-Hall, Inc.)

tuning. The PLL circuit is much more stable than its mechanical tuning counterparts. This system is composed of several blocks. Each block plays an important role in the system. It is difficult to describe the system if one attempts to look at it as an input–output system. The only way to handle a circuit of this type is to start somewhere and describe the entire system. Keep in mind that all the action described here goes on at the same time.

A block diagram for a simple PLL system is shown in Figure 6-12. It has three basic blocks. These are the frequency reference, a voltage-controlled oscillator (VCO) and a phase comparator, or detector block. The VCO is designed to operate at, or near, the desired frequency. A reference oscillator also operates at the desired frequency. Both of these blocks have an output that feeds into the phase comparator block. The output of this block is filtered and becomes a dc control voltage. The dc control voltage is used to bias the VCO. If the VCO frequency should change, the output of the comparator also changes. The change that occurs is designed to bring the VCO back to its desired frequency. The VCO output frequency is constantly monitored in this manner. The PLL circuit keeps it on the desired frequency. A second output from the VCO is connected to the desired blocks, usually a mixer in the tuner.

The system described in the preceding paragraph is not too practical. It is used to show how the system functions. A more practical application of a PLL circuit is shown in Figure 6-13. In this system the VCO and reference oscillator operate on different frequencies. The reference frequency

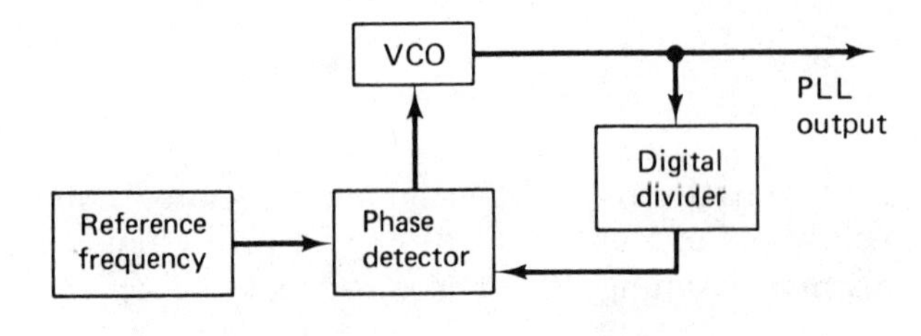

Figure 6-13 Block diagram for a digital divider PLL system. (From Joel Goldberg, *Fundamentals of Television Servicing,* © 1982, page 215. Reprinted by permission of Prentice-Hall, Inc.)

is usually higher than the VCO frequency. One additional block is required for operation. This block is called a *divider*. It divides the VCO frequency by some predetermined amount. If this confuses you, look at it in this manner. When the signal is divided, there are more parts, thus a higher frequency rate. The higher oscillator frequency is often more stable than lower frequencies; therefore, it is used in place of lower-frequency circuits.

The output of the phase comparator is still a dc voltage. The comparator circuit develops the dc output voltage from the comparison of the reference and divided VCO signals just as previously described. It is also applied to the VCO as a correction voltage. The VCO frequency is actually controlled by the correction voltage. The voltage changes the frequency of the VCO and will either move it to another frequency or keep it a preselected frequency.

The system as it is described requires more blocks if it is to work successfully as a variable-frequency tuner. The addition of two more blocks provides the versatility required. These blocks and their position in the system are shown in Figure 6-14. The two blocks added to the system are dividers. One of these is a fixed divider. Its purpose is to divide the reference oscillator signal down to a 1-MHz rate. The second divider is programmable. This means that its rate of division is adjustable. The adjustment may be accomplished by a switching mechanism or by use of a variable-frequency oscillator. In either form the end result produces a wide variety of frequencies from the VCO.

There are advantages to using a PLL system. One of these is that the operating frequency is "locked" to the desired point. Another is that it is possible to use a microprocessor in order to select the desired frequency. A third is that it is reasonably easy to develop a digital frequency display from the electronics in the system.

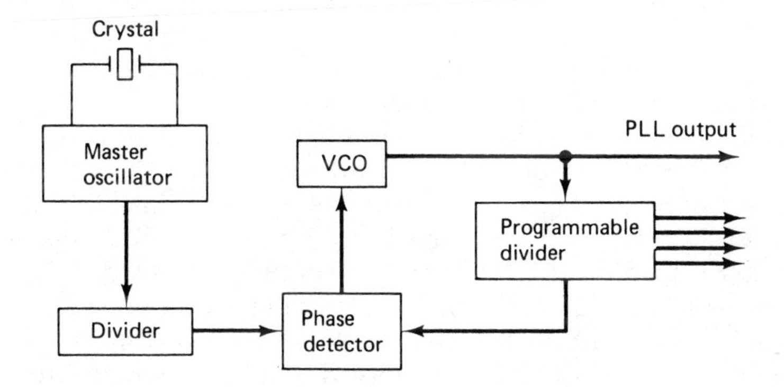

Figure 6-14 Block diagram for a digital divider PLL system with a programmable divider. (From Joel Goldberg, *Fundamentals of Television Servicing*, © 1982, page 215. Reprinted by permission of Prentice-Hall, Inc.)

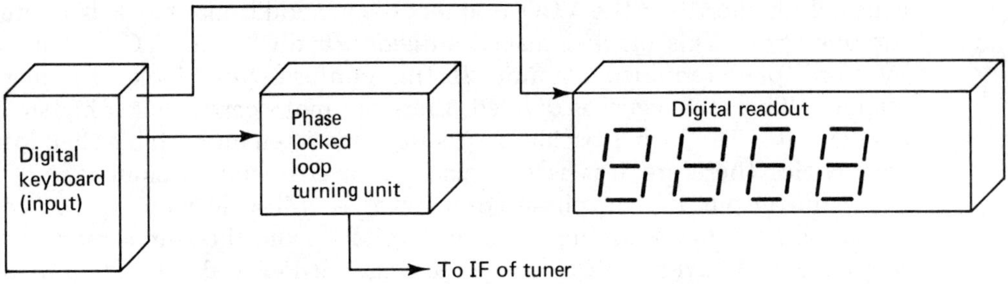

Digital keyboard (input)

Phase locked loop turning unit

Digital readout

To IF of tuner

Figure 6-15 Block diagram of a PLL system using digital frequency readouts.

A block diagram for this system is shown in Figure 6-15. The digital keyboard develops an output voltage. The exact level of dc output voltage depends on the numbers that are punched into the keyboard. Inside the keyboard is a microprocessor unit. It converts the raw data of the desired frequency into a dc output voltage. The level of dc output voltage is directly related to the frequency. For example, a frequency of 88.0 MHz may produce an output voltage of 0.15 V. At the other end, a frequency of 106 MHz may result in a dc voltage of 4.8V. Frequencies between these values will produce voltages between 0.45 and 4.8 V. This dc output voltage is fed into the PLL circuit programmable divider. Its division rate is determined by the value of dc voltage. The divider output is compared with the reference oscillator output. The comparator output changes the VCO frequency so that it corresponds to the frequency that is selected at the keyboard.

Electronic signals created in the keyboard unit are also used to establish a frequency display. A series of digital converters and dividers are used in order to develop a readout. The readout corresponds to the frequency at the keyboard. The system used to develop this display is similar to that used for displays of time or frequency counters. Service of digital display units should be attempted only when one has a good understanding of how these circuits function. This material is beyond the scope of this book. A block diagram for the readout system is shown in Figure 6-16 for reference. The programmable divider or keyboard data are developed as a digital format. This information has to be translated into a decimal information in order to be meaningful to the consumer. A unit called a BCD-to-seven segment converter is used for this purpose. It takes the binary-coded-decimal information and uses it to turn on a combination of up to seven segments of the readout. The proper combination of these segments will display the numbers from 0 to 9. A group of three or more readouts is used in order to display the frequency.

The purpose of any tuner is to develop an audio signal from the modulated RF signal being transmitted. The tuner must be able to select one station from the many being broadcast. It must have sufficient sensitivity to

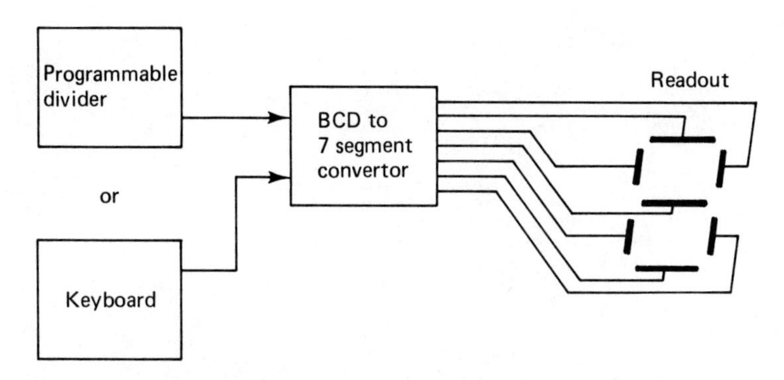

Figure 6-16 Block diagram showing microprocessor and system used to drive a seven-segment LED display.

pick up the weaker signals. The combination of the RF amplifier, mixer, and local oscillator blocks produce an IF signal. The IF block provides about one-half of the total signal amplification provided in the tuner–amplifier package. In addition to this the tuner has to be able to remove, or demodulate, the information found on the station carrier signal. There are times when two sets of demodulators are required. This is true for FM stereo. The end result of all the correct action is an audio signal. It reproduces the modulated information for amplification by the audio amplifier and ultimately as sound waves from the speaker.

QUESTIONS

6-1. What is the difference between a tuner and a receiver?
6-2. What is receiver selectivity?
6-3. What is receiver sensitivity?
6-4. Name the blocks found in an AM tuner.
6-5. Draw a block diagram for an AM tuner.
6-6. What additional blocks are used in the FM tuner?
6-7. Name the tuning systems used in tuners.
6-8. What is a VCO?
6-9. What is the purpose of a phase comparator?
6-10. How does the phase comparator control VCO frequency?

Section II

TROUBLESHOOTING AND REPAIR

This portion of the book is devoted to the specifics of how to repair stereo units. Each chapter discusses one topic or section of the stereo unit. The format for each chapter is first a discussion of the theory of operation for the specific section. The discussion covers circuits that use discrete devices as well as integrated circuits.

Examples of circuits used by manufacturers of modern stereo system components are used wherever possible. Schematic diagrams show these modern circuits. Techniques suggested by these manufacturers are presented. In this way the reader has the opportunity to keep up with the latest servicing techniques.

Chapter 7

Introduction to Troubleshooting

Each person has his or her own method of starting a project. This is also true in the field of stereo servicing. When a set requires repair the customer usually does not know exactly what is wrong with it. The service technician is usually asked to estimate the cost of the repair and the length of time required to fix the set. Realistic estimates can be made only after determining the location of the problem. The technician requires information in order to do this. Successful technicians seek information and try to localize the problem in an orderly manner. The best way for a service technician to handle any set that requires repair is to establish a working system for approaching the repair problem.

Often, many repairers want to open up the set as a first step. The next step this person takes is usually to start to poke around in the back of the set to see what can be found that looks like a trouble area. Many hours of time are wasted with this approach. The goal of a competent electronic repair person is rapid and correct repair. In many service organizations the technician receives a base salary and a commission. Take-home pay is based on successful repair of sets at a rapid rate. Efficiency is important to the servicer if paychecks are to contain large numbers of dollars.

The highly successful technician does some work before opening up the set. This work is required to increase efficiency and productivity. The work plan is firmly established in the technician's mind. Often, a checklist is used to help novices become experts in this field.

The first step in any repair is to talk to the customer. If a customer brings a set to the shop, a technician should greet this person. Questions are asked relating to the sounds, sights, and smells associated with the malfunctioning set. The customer will often help the technician in localizing a service problem. Careful listening to the answers given by the customer is very important. The attitude of the technician toward the customer is also very important. Very often the customer is certain that he will be cheated in some way by the servicer. Too many stories are circulated that purport to show that technicians are dishonest. It is very important for the technician to gain the customer's confidence. This is easier to accomplish when the technician is neatly dressed and clean. Information is obtained from the customer by asking questions related to how the set was operating as it failed. The customer's description is heard. Again, questions are asked as more information is collected. The technician should also attempt to obtain a service-related history of the set from the customer.

Once the technician has listened to what the customer has to say about the operation of the set, the set is plugged into a power source and turned on. At this point the back has not been removed. The technician listens for abnormal sounds in the set. These include frying, popping, arcing, and, of course, a lack of any noises. The technician also carefully looks for signs of smoke and attempts to smell for overheated components.

At this point a first major decision is made. There are two directions in which to go. One is determined by a "dead" set. It has the appearance of not receiving any power. None of the usual operating sounds, sights, or smells are present. The decision to make is: should the back of the set be removed at this time? The answer is *no*! There are other checks to make before removing the back. These checks include the following:

1. Is there a circuit breaker or fuse on the back of the set? Often the circuit breaker opens and all that has to be done is to reset it. In some cases a fuse will fatigue and fail. This, too, may be checked from the back panel of the set before the back is removed.
2. Are the controls in their proper position? Misadjusted customer controls, such as volume and function, may give the appearance of a non-operating set. Adjusting the controls will often resolve what was thought by the customer to be an operational problem.

If after performing these steps, the set still does not operate properly, it is time to remove the back. Very often this is done by the technician after the customer leaves. It may sound foolish, but the person receiving the set for repair should note any missing knobs or other parts, or any scratches on the cabinet. These are called *cosmetic defects.* Where possible, these should be mentioned to the customer when the set is received for repair. This will forestall later arguments or bad feelings.

All parts removed from the set should be placed in a bag. This procedure minimizes the problem of lost parts or fasteners. At some repair facilities, knobs are washed and cabinets are cleaned as a regular part of the repair process. One may purchase cloth bags that have a drawstring closure. These are ideal for holding knobs and fasteners. The bag string is tied to the set to keep all parts together.

Another point to remember is the cosmetic appearance of the set. Cabinets may become scratched if one does not take care. A piece of clean carpeting is placed on the work surface before the set is placed there. The carpeting protects the exterior of the set from damage.

When the set back is removed, a replacement power-line cord is used to provide power for the set. The wise technician has an isolation transformer on the workbench. This transformer provides electrical isolation between the set and other electrical equipment. In many sets one side of the power line is connected to the chassis. It is possible to receive an electrical shock unless the chassis is isolated in this manner. Connections are made to the antenna as well as to the power source. The set is turned on again. If the set works at this point, the power cord is probably defective. Replacing it will repair the set. If this is not the problem, further repair is needed.

At this point the technician has to make another major decision. This is based on the information collected so far. The question to be answered is: Is the set totally dead? If the answer is "yes," certain key parts of the set are located and checks are made. If the answer is "no," other types of checks are made. Each of these checks is made based on the knowledge of how any set functions. Service information must be available for use by the technician. Information related to block diagrams and signal paths is also important required knowledge for the technician.

DEAD SET

A dead set indicates a power supply problem. If there are any tubes in the set, their filaments are not glowing. No sound is heard in the speaker. This also includes a low-level 60- or 120-Hz hum. The presence of hum usually indicates that the power supply section is working. If it is determined that the set is dead, the technician follows repair techniques described in Chapter 9.

If the tube filaments light and one can hear some sound in the speaker, other types of decisions have to be made. At this point the technician looks for broken wires, charred parts, broken tubes or components, or any other visual signs of defects. Some people prefer to do this as a first step before the set is turned on. In either situation if smoke signals appear, one will usually find burned or charred components or wiring. Smoke signals often help localize a problem. Usually, the identification of the component will relate it to a specific circuit in the set.

If the set is not completely dead, it is possible to localize problems by observing and listening. This process will help localize the trouble area to

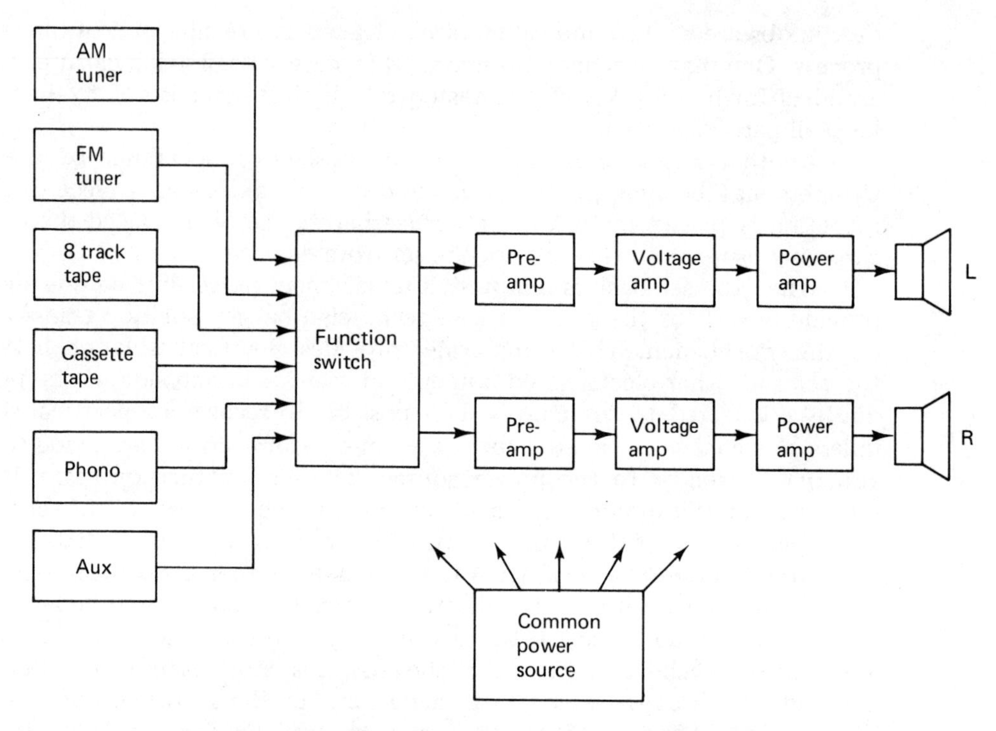

Figure 7-1 Block diagram of a stereo receiver system.

one major section of the receiver. Figure 7-1 identifies major areas of the receiver and relates these areas to the appropriate blocks. Use this information as you read the following section.

SYMPTOM DIAGNOSIS

Troubleshooting is an art. When it is done successfully, the results are gratifying. The approach used by the service technician can make a repair a burden or a pleasure. The idea is to use all of your senses in order to start the repair in the correct manner. This approach works well for all types of repair and service, not just for stereo equipment. The initial step is to analyze the symptoms.

The qualified repair technician knows what to expect to find in a unit before troubleshooting begins. The specific wiring and parts layout need not be known at this time. Turning the unit on will provide some initial information. One of the first things to look for is some sign of operating power. This sign could be that the dial lights light or that some sort of "on" indicator is functioning. The fact that these indicators are working shows that some power is available for the unit. All that this proves at this point is that the power-line cord and a part of the power supply, the power switch, and fuses or circuit breakers are functioning.

The next thing to do is to listen. When power is applied there should be a "plip" sound from the speakers. This sound indicates that the power supply for the output stages and the output stages are working. One should listen to be certain that both channels of the stereo unit are producing this sound. Another listening sound is the background noise associated with the volume control. The rotation of this control often increases background noise or hum. These sounds indicate that the section of the unit located between the volume or loudness control and the speakers is working.

Still another check is to rotate the function switch. If some of the inputs are working properly, the problem is isolated to those inputs that do not work. A wire or screwdriver blade placed on the input connector should produce some hum or noise in the speakers. This will provide a quick audio check.

In cases such as those described previously, the area of the problem is reduced before the back of the set is removed. The successful technician uses this type of information to speed the correct repair. This does not mean that the set can be repaired without further tests. The purpose of this analysis is to localize a trouble area. One of the most difficult things to impress on the beginning technician is that efficiency does not always require physical activity as a first step. The more successful technician spends the initial portion of the repair time in analysis of the problem. In these situations thinking must precede action. The tendency is for a person to "jump in" and start taking things apart or to replace components in the receiver. These are often required actions, but they are *not* the first steps to be taken.

Analysis of the problem requires a certain amount of time. Analysis requires information from the receiver and the customer. This information is evaluated together with information obtained from the service literature for the receiver. These inputs help the technician develop a plan of work for the repair. The plan of work requires knowledge of signals, signal paths, expected waveforms, and how operating power is obtained and distributed in the set. This approach is valid as well for other appliances, automobiles, or any other type of device. The technique of planning based on analysis will help develop proper repair procedures.

OPERATING POWER CIRCUITS

Problems created by operating power circuits must be considered during the repair process. There are two types of problems that may occur. One of these is related to the failure of a filter capacitor. This may cause a low operating voltage. It may also produce a 60- or a 120-Hz wave in a section of the set. If this occurs in the audio section, the result is a low-frequency hum that is reproduced and heard at the speaker.

Another type of filter capacitor failure allows signals from one circuit to enter a second circuit by way of the power supply. Filter capacitors are actually "decoupling" capacitors. Their purpose is to remove any variations in operating voltage. Signals are variations in operating voltage in many circuits. In the collector circuit of a transistor this is desirable. The variations are then coupled to the next stage. This variation in operational voltage in one stage must not be transmitted to any other stage through the power source. Capacitors C_1 and C_2 in Figure 7-2 are used as decoupling capacitors. Any variations in operating voltage caused by the signal are decoupled, or bypassed, to common by capacitor action. A second type of operating power failure is caused by a component failure. This failure requires dc circuit analysis and test equipment.

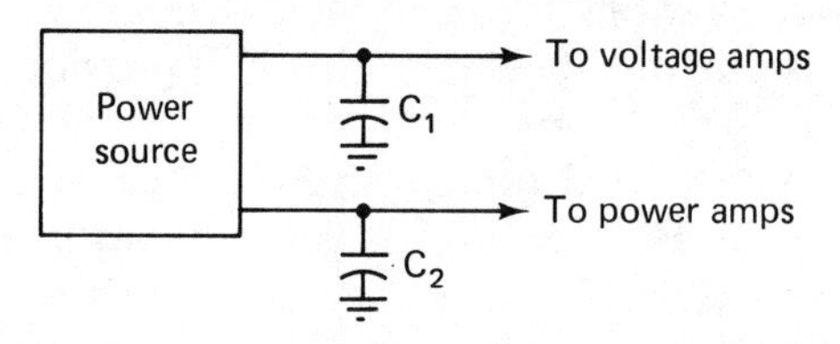

Figure 7-2 Capacitors C_1 and C_2 are used to "decouple" signals between stages and the power source.

At this point you will realize that the successful technician requires much information. This information is needed throughout the repair process. Decisions must be made at each step of the repair. Information is required for valid decision making. The next three chapters provide further types of information required for analysis of the problem.

Procedures outlined in the preceding section provide a broad background for troubleshooting. Once these procedures are developed, the technician is ready to start the repair. In this process the technician makes tests and measurements to locate a specific problem or defective component. The technician starts by suspecting that the whole set may be bad. Using a well-developed system reduces the suspected trouble area. This may require several steps and measurements. Each step further reduces the area by eliminating properly working sections until only a small section of the receiver is left. If this procedure is followed correctly, the problem is then isolated.

There are several procedures that can be used when locating trouble areas. One method starts at the signal input of the receiver. Signals are traced through the set until they disappear. When this occurs the trouble area is located. It is between the last two test points. A second method is the reverse of the one just described. Tests are started at the output of the set. The procedure then is to work back toward the input point. When the signal is located, the area of trouble is also located.

Both of these methods are slow. They are time consuming and inefficient. There is a much better way to troubleshoot any stereo system. Good use of analysis by sight, sound, and smell helps very much. This procedure should be developed until the technician is expert at it. It may be further refined as set operation knowledge increases. The next phase of troubleshooting relates to a system for analysis. This system may be applied to signal paths and to operational voltages.

The analysis procedure uses five basic systems. These are the only systems that are used in electronic devices. They should be learned. Recognition of each is important to the analyst. Each of the five systems is discussed in this chapter. The information is then related to typical receiver blocks and circuits. Use of the proper system at the appropriate time will raise the effectiveness of the technician.

LINEAR PATHS

A linear device is one that is in one line. Such a system is shown in Figure 7-3. Input to the system is on the left-hand side. There is only one output. It is on the right side of the diagram. Signals are processed through each block from the left to the right. The signals may change in size or shape as they are processed. Service literature for a specific receiver will provide specific information related to size and shape. The basic fact is that there is only one path for movement in a linear system. Signal flow should not be interrupted.

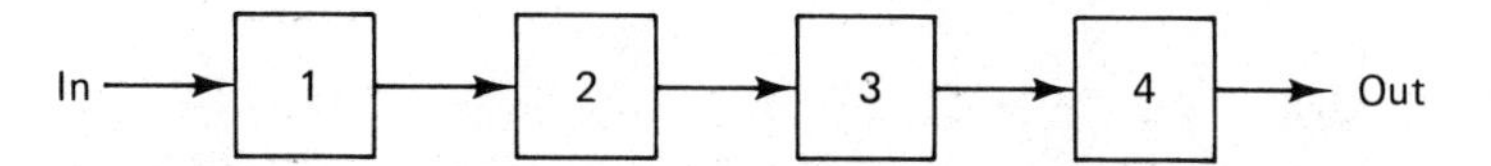

Figure 7-3 Block diagram for a linear-signal-flow-path system.

The procedure for troubleshooting the linear system is as follows:

1. Be sure that the input signal is present.
2. Check the output. If proper signal is present, the whole section is working properly. If there is an incorrect output, or no output, the system is not working properly.
3. The next check is to be made at or near the middle of the path. This point is located between blocks 2 and 3 in the illustration. This procedure is shown in Figure 5-2. The results of the test will provide information for the next step.
4. Evaluate the test results. Figure 7-4(a) shows a set of brackets at each end of the blocks. This is the original area of trouble. It has to be reduced to a very small portion of the system. One of two results will be found at the midpoint of the system. Either signal will be present or it will not be present. One of the brackets is moved depending on the test results. If signal is present at the midpoint, blocks A and B

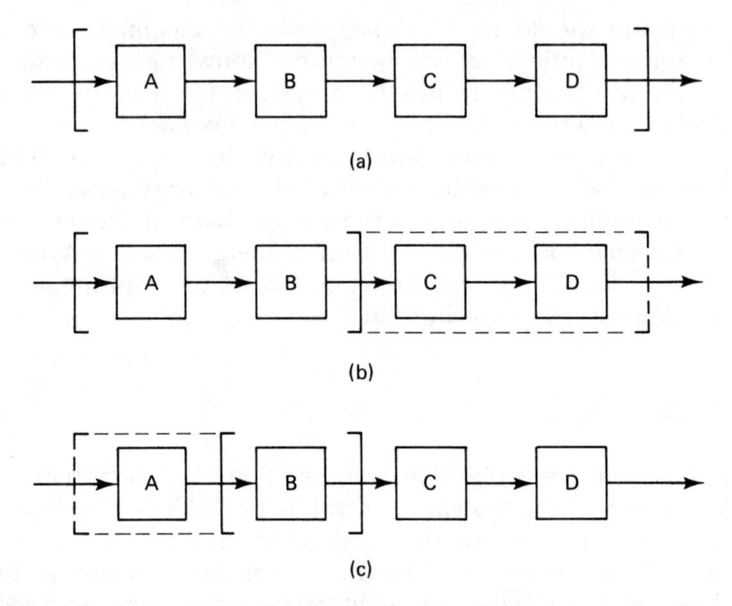

Figure 7-4 Method of isolating troubles in a linear system. Results of tests made at the midpoint of the system reduce the brackets to an area of trouble.

are working properly. If no signal is present, blocks C and D are working properly.

5. Move one of the two brackets around the system to the midpoint. In the example the right-hand bracket is moved because no signal was found at the midpoint. This is shown in part (b) of the illustration.
6. Make another test at, or near, the midpoint of the remaining blocks. The evaluation at this point is the same as that performed in step 4. In the sample a signal is found at the test point. The area of trouble is then considered to be in block B. The left-hand bracket is moved to this block. The brackets now surround the one block in which trouble is located. The next step is to locate a specific component that has failed. Procedures for doing this are discussed later.

This procedure is easily related to the signal path in a stereo system. A schematic for a phono preamplifier is shown in Figure 7-5. The signal path for this circuit is a linear path. The signal input is at R_{117} and C_{101} to the base of Q_{101}. The output element for this transistor is its collector. It is directly coupled to the base of Q_{102}. This transistor is wired as a common-collector amplifier. Its output is at its emitter. The emitter is directly coupled to the base of Q_{103}. Its output element is at the collector. The signal is then coupled through C_{108} and C_{109} to the next stage.

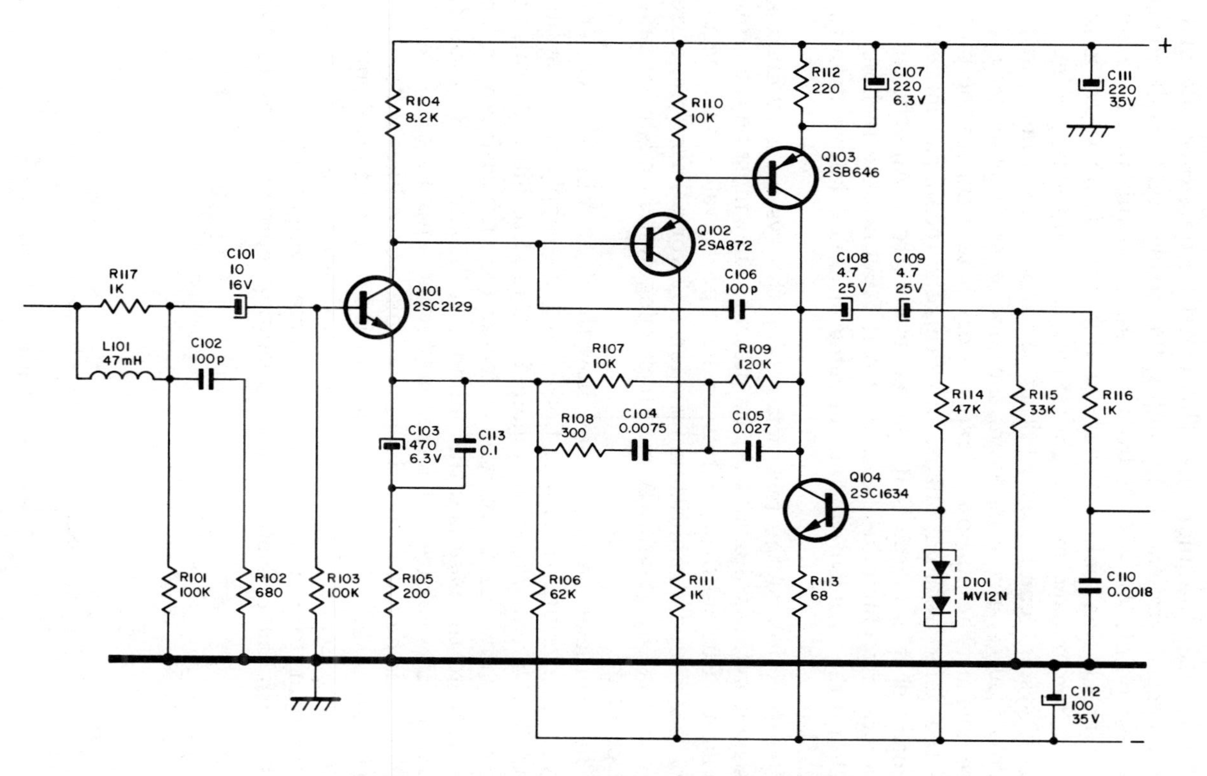

Figure 7-5 Linear-signal-flow-system schematic diagram.

The first check for a system of this type is at the input. Once the signal is found at this point, the output is checked. A lack of signal indicates a failure between these two points. The system is now split in half and a check made at or near the midpoint. This could be on either side of transistor Q_{102}. If signal is present at this point, the bracket on the left side is moved to the test point. All is well from the input to this point. There is little need to check this section further at this time. If no signal is present at this point, the bracket on the right side is moved to the test point.

This system eliminates working sections by constantly reducing the suspected trouble area. Eliminating the good areas increases the efficiency of the technician. The purpose of repair work is to diagnose the problem and locate the defective component. After this the technician makes the repair and finally checks to see that the repair is correct. An efficient system of troubleshooting eliminates wasted effort. It allows the technician to concentrate on the problem without bringing in useless information to confuse the problem.

One function of the technician using this system is to make a circuit analysis. Operating voltages must be present. They must be the proper value and polarity. Current flow, and its resultant power value, must occur if work is to be performed. Amplification is a form of work. Therefore, current flow must occur in the system. A malfunctioning system might be caused by a power problem.

The technician needs to know how the devices in the receiver operate. The system described as a linear flow system also applies to operating power circuits. The circuit shown in Figure 7-6 is used to illustrate this principle. The circuit shown is a common-collector (or emitter-follower) amplifier. The input is to the base and the output is at the emitter. Resistor R_E establishes the emitter voltage and is the load for the output signal.

When this circuit is functional, the collector voltage is 23.0 V. Emitter voltage is about 2.0 V. This information is found on the schematic diagram included with service literature. The current-carrying components in the output circuit form a linear circuit. These components are R_E and the transistor emitter–collector leads. Operating current flows in this linear

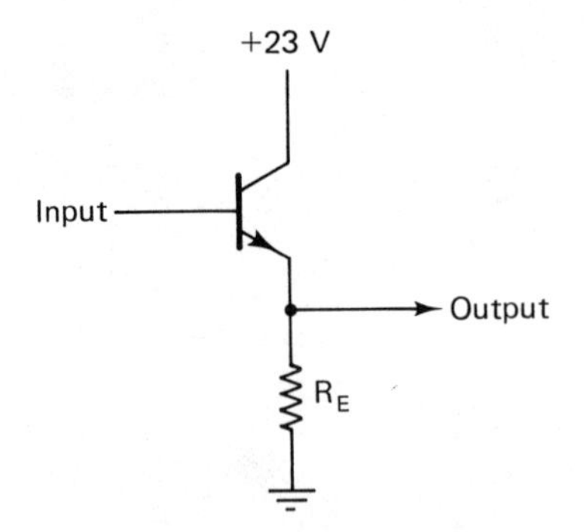

Figure 7-6 Emitter-follower amplifier circuit used as an example in the text.

path. Voltage drops develop across each component. The action of the current flow produces the operational voltages shown at each point on the diagram.

Tests performed by the technician show that the input signal at the transistor base is good. There is no output signal. The immediate thought would be to replace the transistor. This may cure the problem, *but* it also may not cure the problem. Other tests should be made before any components are changed.

The first step is to check the input of the operating voltage. This system is also a linear system. The steps used for testing a signal processing system also apply for this circuit. If input voltage (+23.0 V) is correct, a second test is made at, or near, the midpoint. This is at the emitter of the transistor. Information given on the schematic shows 2.0 V at this point. If this voltage is incorrect, the reason has to be determined.

A measurement of 0.0 V at this point usually indicates an open circuit between the test point and the power source. The transistor is open. A measurement of 23.0 V at this point indicates a no-current condition. Voltage drops occur only when current flows. No voltage drop indicates that no current is flowing. This is an open circuit. This opening is between the test points. One meter lead is at circuit common. The second lead is at the emitter of the transistor. The opening is between these points. A resistance measurement will help to determine which component has failed.

Once the specific defective component is identified, it is replaced. This will restore the circuit to its operational condition. The procedure for troubleshooting a linear current path is the same as that for troubleshooting a signal path. The difference is in the measuring equipment. In either type of circuit a schematic diagram is used to supply normal operating circuit values.

SPLITTING PATHS

The splitting-path system is used to separate two or more signals. It is also used to send the same signal to two different circuits or components. A schematic for a splitting circuit is shown in Figure 7-7. An audio signal is presented to the base of the input transistor Q_1. This transistor is operated as a phase splitter. It has two outputs. One of these is from the collector. It is connected by means of a capacitor to the base of one of the output transistors Q_2. The second output is capacitive coupled from the emitter of Q_1 to the base of Q_3.

This system is checked at the point of, or component, where the split occurs. The procedure for troubleshooting a splitting-path system is as follows:

1. Check to be sure that the input is present.

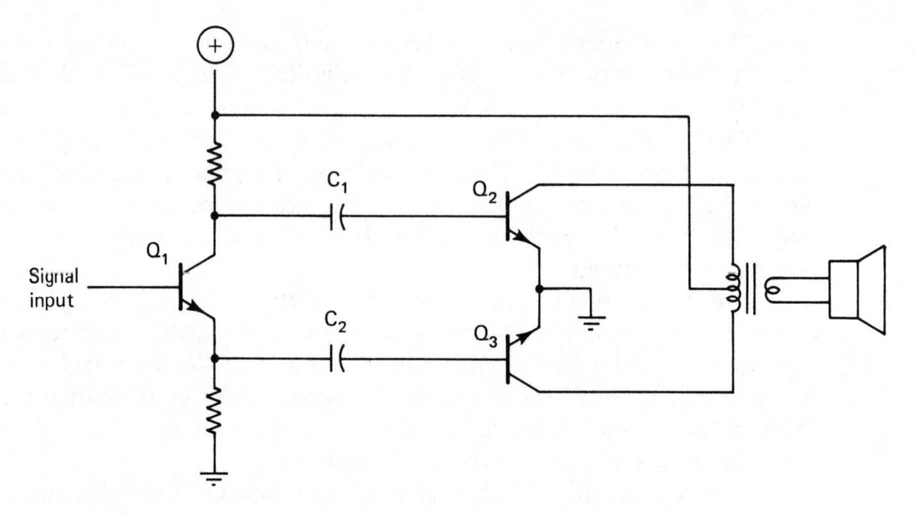

Figure 7-7 Schematic diagram for a signal-splitting circuit. The output of Q_1 is split into two paths in this system.

2. Check both outputs.
3. Evaluate the test results. If both outputs are correct, the unit is good. If only one output is working, the problem is most likely to be in the coupling capacitors C_1 or C_2. It could, however, still be in transistor Q_1. The effectiveness of this system is its adaptability. The technician is able to apply this approach when first evaluating the stereo unit. The presence of undistorted audio indicates that the system is functioning. If the amplifier has no audio output, the trouble could well be after the split. This sytem could also be first tested as a linear system. The initial test could be made at the input to transistor Q_1. This will quickly help to determine if the problem is before or after the point of separation.

The approach used to check signal path systems is also able to be used for power source systems. Many of the power supplies found in receivers use a separating current-flow-path system. One such system is shown in Figure 7-8. There is one output from the power supply block. Current paths are separated after this output in order to supply four systems. If all systems were not functioning, the most logical test point is at the point of separation. A measurement at this point moves one of the brackets. If voltage is present at this point, the bracket on the left side is moved to the test point. The power source block is good. If there is no output at this test point, the power source block is not functional and the right-hand bracket is moved.

If the input point is good, the outputs are checked. If one of the outputs is not present, the trouble area is located between the point of separation and that output point. The right-hand bracket is moved to the

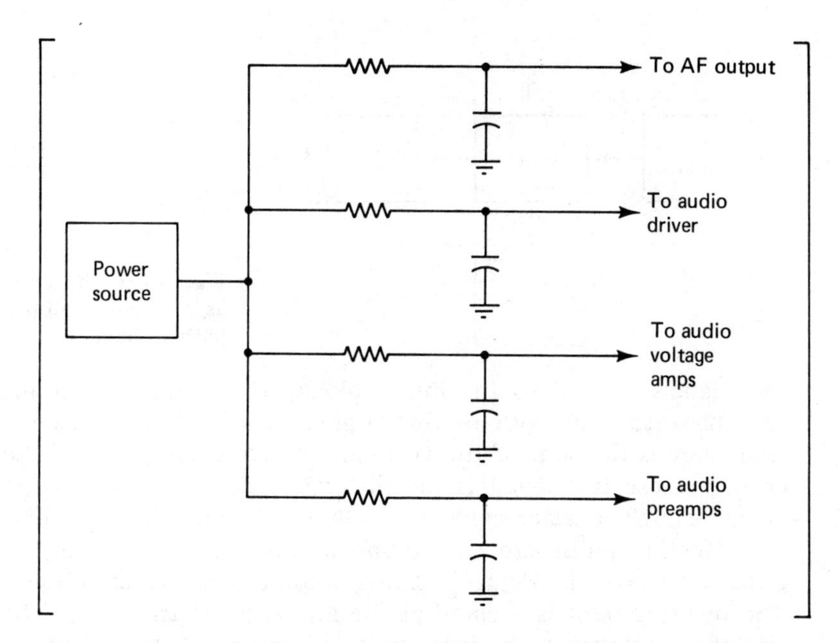

Figure 7-8 A splitting system may also be used for operating power distribution, as shown here.

right of that specific output. This portion of the system is treated as a linear path system for troubleshooting purposes.

MEETING PATHS

This system is the opposite of the separating path system. Two or more signals enter a junction. Some sort of electronic interaction occurs. There is one output from the junction, or meeting, block. The procedure used to troubleshoot a system such as this is as follows:

1. Check at each input point to see if there is an input signal with an oscilloscope.
2. Check at the output to determine if it is functional with an oscilloscope.
3. If both inputs are working and the output is incorrect, the trouble is in the common block.

The block diagram shown in Figure 7-9 is a classic example of this system. It is the tuner system. Two inputs to the mixer block are required. One of these is the modulated RF carrier signal from the transmitting station. The second is the unmodulated carrier signal from the local oscillator. These

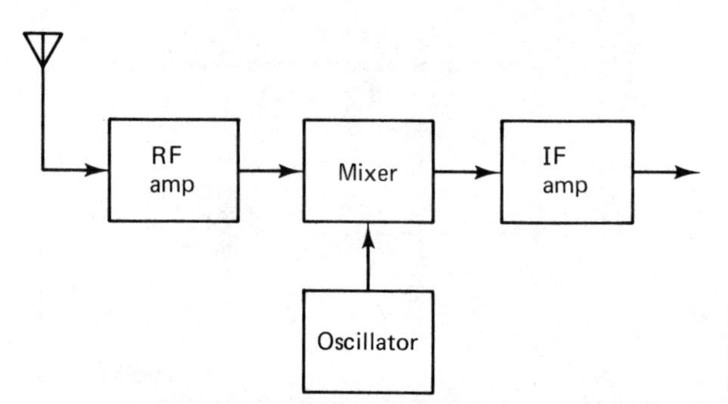

Figure 7-9 The heterodyne tuner is an excellent example of a meeting-path system.

two signals are fed to the mixer block. The output of the mixer contains four basic signals: both of the original signals, a modulated carrier whose frequency is the sum of the two carrier frequencies, and another modulated carrier whose frequency is the difference between the two original carrier frequencies. The latter signal is usually the IF amplifier frequency.

Meeting paths are also found in electronic power circuits. One such circuit is shown in Figure 7-10. A common-emitter amplifier is illustrated. The meeting path is located at the input to the transistor. This is the base circuit. Resistor R_B, the base–emitter junction of the transistor, and resistor R_E form a current path. This establishes a bias of 4.0 V at the base. A signal source of 2.0 V p-p is also connected to the base. When both bias and the signal voltage are present, the base voltage will swing from 3.0 to 5.0 V. This produces an output voltage swing of 10 V p-p. If both input voltages are present and there is no output voltage, the transistor circuit is malfunctioning. Both inputs must be present to obtain the two outputs. Checking this circuit is done in the same manner as the meeting type of circuit. If both inputs are proper and either one or both outputs are missing or incorrect, the problem is in the filter unit.

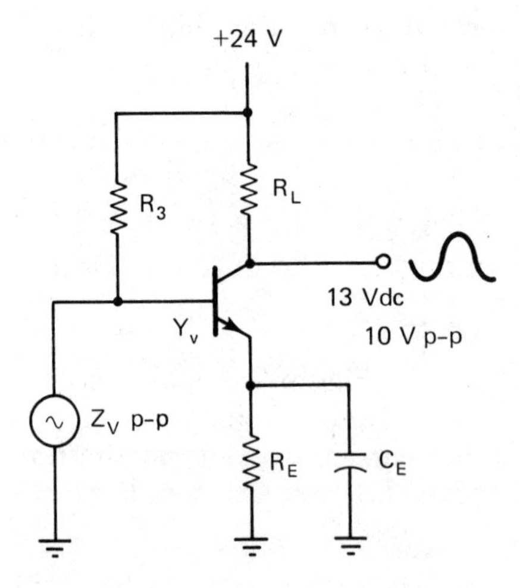

Figure 7-10 Meeting paths are also used for operating power. In this circuit the bias and signal both meet at the base of the transistor.

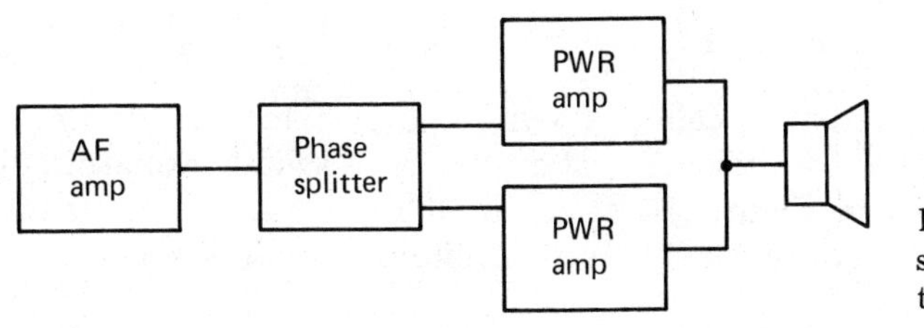

Figure 7-11 This power amplifier system is an example of a separating and meeting system for signals.

A combination meeting–separating circuit is shown in Figure 7-11. This is the phase splitter and AF power output circuit. When an audio signal is present at the phase splitter, both power amplifiers are working. The audio output at the speaker is the result of the combination of signal from each power amplifier.

FEEDBACK PATHS

Automatic control circuits have been found in receivers many years. One of the first types of control circuits is the automatic gain control used in the IF section. The term *gain* is used to show that signal amplification occurs. Most of the automatic circuit controls use a sample of the output signal to control the amount of signal gain that occurs. The AGC block functions in this manner. A portion of the output signal is fed back to the input. This sample signal is used to control the bias on a stage of the IF amplifier. Figure 7-12 shows this.

The procedure for troubleshooting any feedback type of system is as follows:

1. Open the feedback path. If normal levels of signal are restored at the output, the problem is in the feedback circuit. Often, a technician will apply a dc voltage from a secondary source to the AGC circuit instead of disconnecting components in the set.
2. If the circuit is still not operable, the problem is not in the feedback path. It is probably in the forward path. In this case the forward path is the IF amplifier and detector blocks.

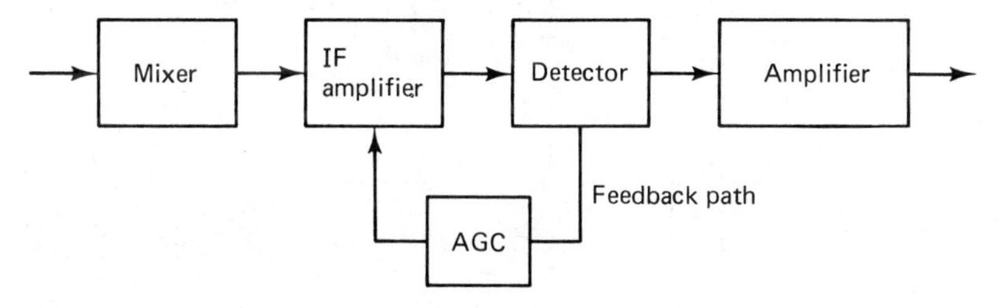

Figure 7-12 AGC systems use feedback principles for their operation.

The final signal path system is called a switching system. In this type of circuit signals are directed by a mechanical or electrical switch. Some circuits, such as the one shown in Figure 7-13, are used to select one input to an audio amplifier. The procedure for this type of circuit is as follows:

1. Use the switch to check each output.
2. If all outputs are functioning, there is no problem. If all outputs are not working, the problem is a block that is common. The only block that is common is the audio-amplifier block.
3. If one output is not proper, the problem is in that circuit.

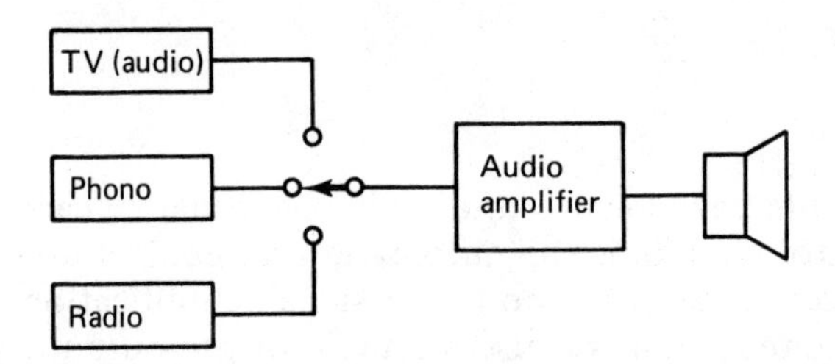

Figure 7-13 Switching systems use mechanical switching to select the desired input.

This approach may also be used for power circuits. A switchable power supply circuit is shown in Figure 7-14. If all three of the output blocks do not work, the problem is going to be in the power supply or its switch. If one of the three output blocks does not work properly, the problem is in that block.

Each of the five signal path systems is used extensively in stereo systems. The technician's role is to be able to identify the system. In most instances this may easily be done by use of a schematic diagram. Once the type of system is identified, the technician follows the procedures

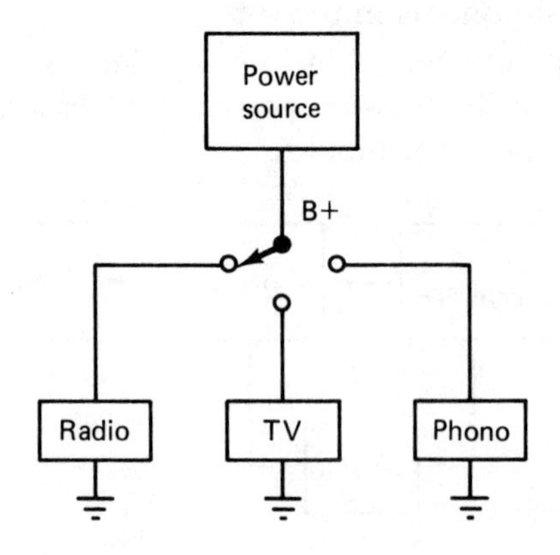

Figure 7-14 Switching is also often used to provide operating power to a specific input system.

explained in this chapter. This will seem slow at first, but do not give up or try to bypass part of the system. Diagnostic time will decrease as experience is gained. After awhile, use of this system will increase effectiveness. It also has the advantage of allowing a person to keep up with technological changes. Repair procedures change as components change. The diagnostic system does not change.

INTEGRATED CIRCUITS

One of the major advancements in electronics was the introduction of the transistor and its solid-state technology. A second advancement was the utilization of the transistor in the integrated circuit. These components have simplified set construction. Reliability of set operation has also increased. Production costs are greatly reduced. More complex circuitry is used in today's receiver than was found in earlier production receivers.

These improvements have great significance for the technician. It means that troubleshooting and repair may be faster and easier. This is true only when the technician is willing to think with an "IC mind." Servicing techniques have changed. The successful technician changes thinking to keep up with these changes. It is important to know how to use test equipment. In addition, one must know which piece of test equipment to use for a specific job. Signal path analysis is easy when one is dealing with integrated circuits. Signal(s) at the input to the IC are checked. The signal(s) at the output are checked. The procedures for each of the five basic systems are applied as required. Operating voltages are also checked. Decisions based on this analysis are made. In many instances the result is the replacement of the IC. It is difficult to accept that this is really faster and easier than attempting to repair a set that uses discrete components.

A commonly used IC is illustrated in Figure 7-15. Most manufacturers show a block diagram of the circuits in the chip rather than the complete circuit. Both are shown here so that one can see the complexity of the circuit. The procedure for checking this circuit is to treat it as a linear circuit. The input is checked and then the system is split. The only coupling device that is not in the IC is one coupling capacitor. This point is checked to determine if the capacitor failed or if the failure is in the IC. Repair is based on the results of the tests.

TROUBLESHOOTING PHILOSOPHY

A certain amount of understanding is required when attempting to repair any receiver. Many technicians have been totally frustrated because they

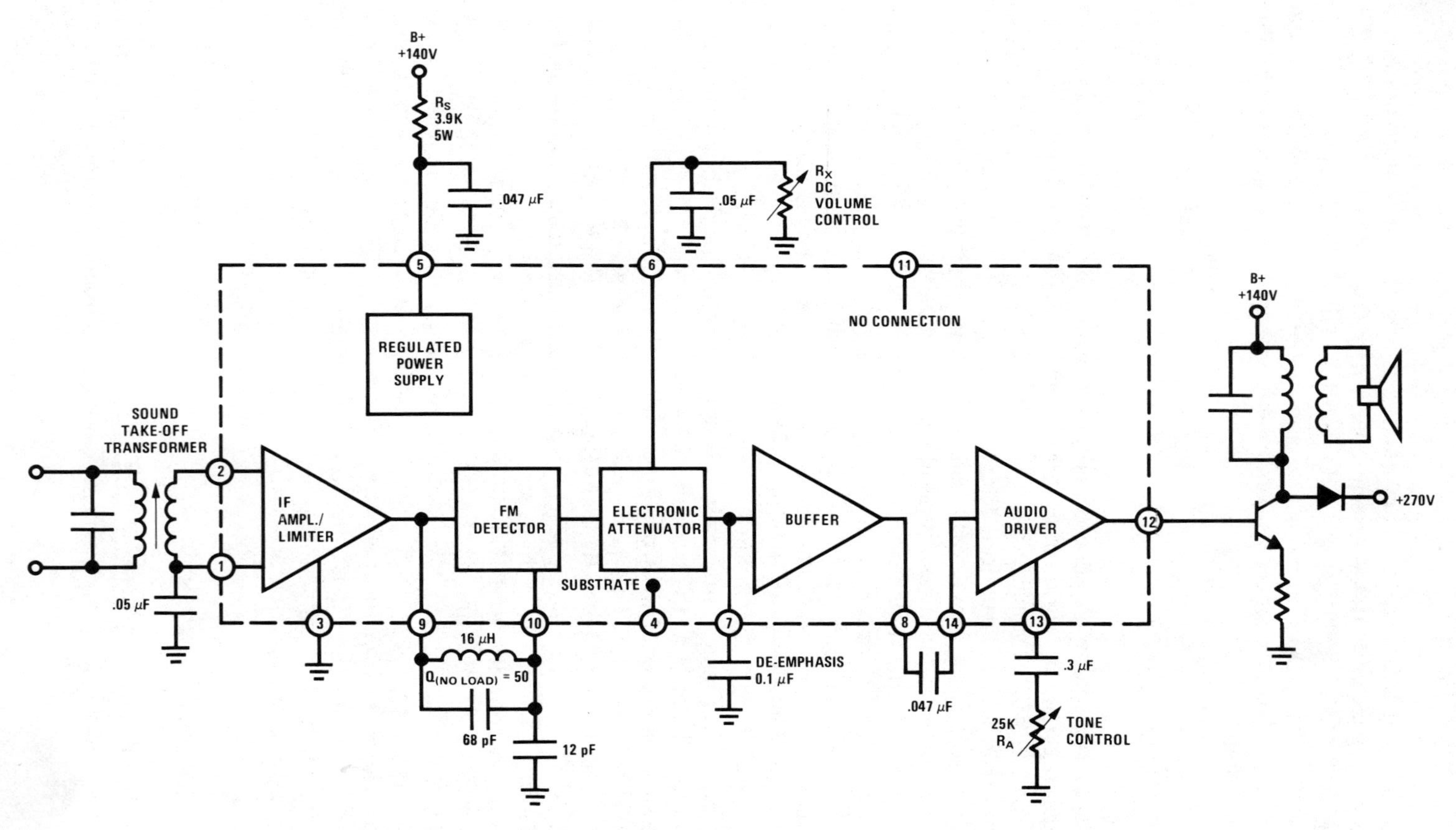

Figure 7-15 Technical information for an IC. Both schematic and block diagrams are provided. (Courtesy National Semiconductor Corporation.)

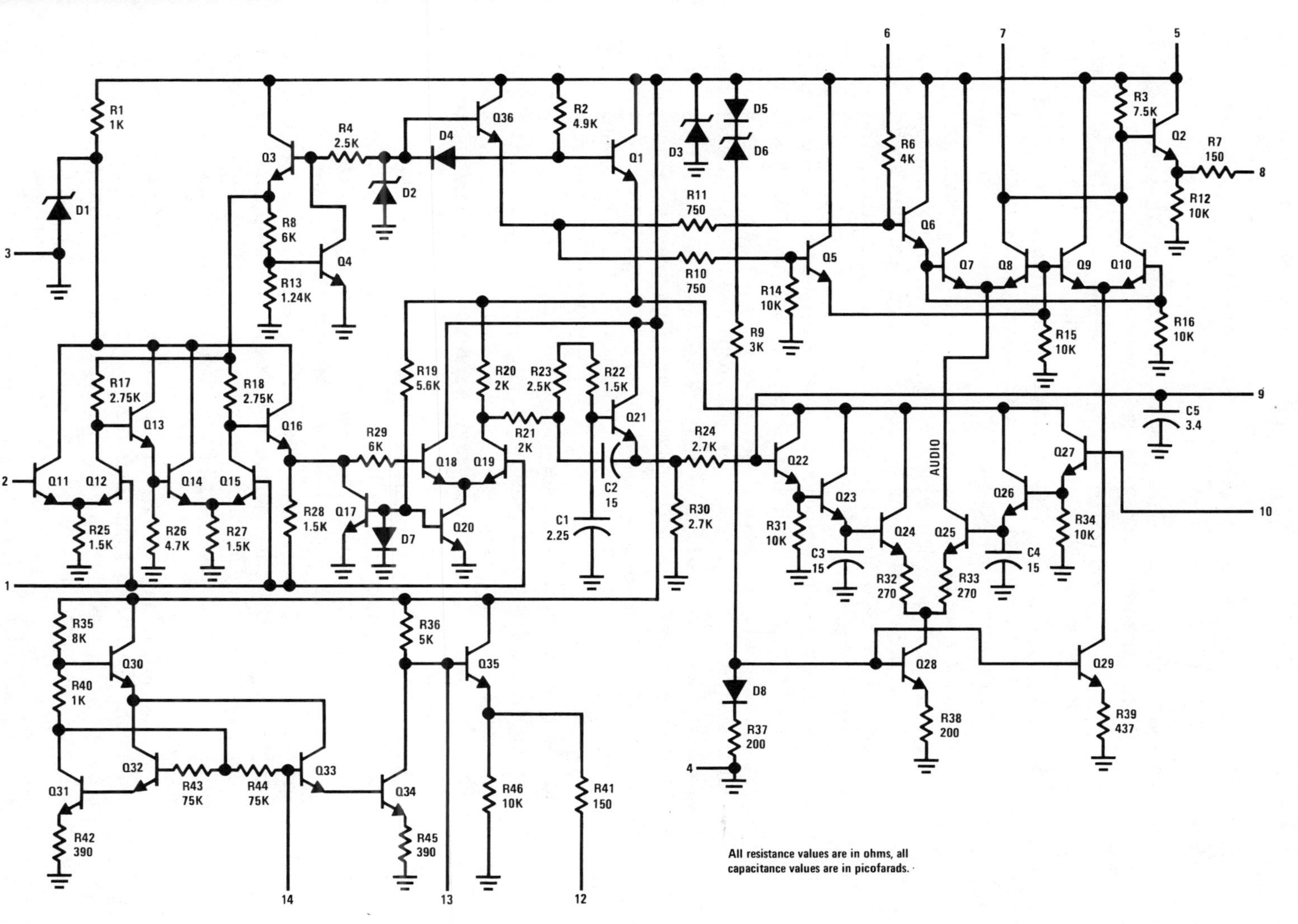

Figure 7-15 Continued.

forget to follow basic repair procedures. The philosophy promoted in this book is as follows:

1. The receiver is an inanimate object. It does not have a mind of its own. It cannot think. Above all, it is not "out to get you" by being devious or sneaky. Keep this in mind no matter how frustrated you may become when resolving a circuit problem.
2. The set did work properly at one time. The manufacturer tested it at the end of the production line. Something occurred either when it was delivered or after it was installed.
3. You, the repair technician, are not being asked to redesign the circuits in the receiver. Design is a function of engineering. Your purpose is to repair a previously functioning receiver. There are times when field modification is necessary. Do this only when instructed to do so by the set manufacturer.
4. Diagnose the problem by eliminating working sections of the receiver. Suspect a large area of trouble as an initial step. Use controls to aid in diagnosis. Reduce the area of suspected trouble in an orderly, professional manner. Keep on reducing the suspected area until one circuit is left. The trouble is then located in this circuit.
5. Repair the receiver. Replace only those components that have failed. The concept of "shotgunning," replacing all components in a stage, is wasteful and can be expensive. In addition, one will occasionally replace parts incorrectly when using this approach. This introduces an additional problem rather than repairing the original one.
6. Assume nothing. Assumptions are often based on inadequate information or on a previous experience. It is true that some defects repeat in the same model of a set. Analyze data and use the analysis to locate an area of trouble.
7. Play the odds. In most instances the odds are correct. For example, hard-working components fail more often than components that do not work hard. Output transistors are hard-working devices. They will fail more often than transistors used as preamplifiers. Therefore, the odds are that output transistor circuits are where the set failure is to be found. A study of service calls as they relate to component failure was conducted by General Electric for their own receivers. The findings indicated that output devices fail more often than other components in a receiver.

   This philosophy is true for other components. Low-wattage resistors fail less often than their higher-wattage counterparts. As you develop skill in troubleshooting and repair, a very definite ranking of the types of parts that fail will be apparent. Do not misunderstand this section. All types of parts fail at some time. The idea presented here is to help you look for the most obvious failures first.

8. Test the repair. No receiver should be returned to the customer until it is bench tested. A good bench test includes at least 2 hours of operational time. Make sure that all troubles are repaired and that new ones do not turn up. Recalls are expensive in time and in lost income. They will occur. Your position is to keep them to a minimum.
9. Last, but not least, charge for your labor. Do not sell yourself too cheaply. Labor costs are high. Equipment, advertising, telephones, and rent are also at high levels. The technician has to earn enough to pay the bills, pay wages, and make a profit. Also, funds should be put aside for replacement of equipment. Electronic service is a professional activity. Highly trained people are employed. Their training has to be kept current. Time spent at a training session is nonproductive service time, but is very necessary. Service income must be adjusted to compensate for this.

The development of a good set of repair techniques will go a long way toward making one successful. The procedures discussed in this chapter provide excellent guidelines. Follow them down the road to success.

QUESTIONS

7-1. What is the first step to be done when troubleshooting a unit?
7-2. What should the technician look for when the set has power applied to it?
7-3. A hum in the speaker indicates that which section of the set is working?
7-4. What sound shows that the audio output section is working properly?
7-5. What is meant by signal tracing, and how is it done?
7-6. Name the five basic signal path systems.
7-7. Name the blocks used in a set that relate to each of the signal path systems.
7-8. Can the procedures used for signal path analysis also be used for operating power? Explain your answer.
7-9. Briefly describe how to troubleshoot an IC system.
7-10. Should the technician redesign circuits? Explain your answer.

Chapter 8

Test Equipment and Its Utilization

The amount of test and measuring equipment required for a stereo service bench depends entirely on the level of servicing and the money available to purchase test equipment. A comparison of the cost of a piece of equipment and the income that will be derived from it plays an extremely important role in equipping the service bench. If one is planning to do general servicing, the equipment will be very basic. On the other hand, if one desires to service some of the exotic equipment currently on the market and to restore this equipment to factory specification, a very extensive equipment list is required.

A fully equipped service bench is very impressive and very costly. An investment of between $5000 and $10,000 is realistic. Some of the equipment is very exotic. Its necessity for basic repair is questionable. Material in this chapter is related to selection and utilization of electronic test equipment as well as a review of basic troubleshooting concepts. Chapter 7 provided a general philosophy for troubleshooting.

BASIC TEST EQUIPMENT

There are some very basic pieces of test equipment that must be available and *used* in the service shop. These include an oscilloscope, a volt-ohm-milliameter, or its electronic equivalent, a power supply, a variable auto-transformer, a frequency counter, and appropriate signal generators. The complexity of today's stereo equipment makes it imperative that good-quality test instruments be used. In general, the following considerations should be given before purchasing any piece of test equipment.

Volt-ohm-milliameter. A meter with a high input sensitivity (ohms per volt) rating is required so that it does not load a circuit when making measurements. A digital readout meter, shown in Figure 8-1, is nice to own, but an analog meter display will function just as well. Solid-state circuits use some very low operating voltages and resistances. The ranges available should permit the measurement of these low values. The selection of a portable or ac-operated instrument is left to the choice of the service technician.

Figure 8-1 A digital voltmeter is often used for measurements because of its accuracy. (Courtesy Leader Instrument Corporation.)

Oscilloscope. It is almost impossible to repair a modern electronic device correctly and efficiently without observing the waveforms used to make it work. The day of beating on a chassis with a hammer or screwdriver to find a defective component is long gone. The qualified technician of today measures and compares voltages and signal waveforms found in the unit to those provided by the manufacturer of the device. The use of an oscilloscope is a fundamental requirement for doing this. A single-trace scope will perform adequately for many waveform measurements. The recommendation regarding an oscilloscope is to purchase one that is capable of displaying two traces at the same time. These are called dual-trace oscilloscopes.

DUAL-TRACE OSCILLOSCOPES

The introduction of digital counters, phase shifters, and delay lines has made the use of dual-trace scope almost a necessity. There are several excellent such scopes on the market today. It is up to the individual technician to select the specific unit according to personal choice and the amount of money he or she wishes to spend. Most of these scopes use the letters A and B to identify the two channels of display. A dual-trace scope uses an electronic switching circuit in order to develop a dual display on the face of the cathode ray tube. There are two vertical inputs and two vertical attenuators on these scopes. Most of them have one time-base generator. It provides a constant sweep rate for both input signals. Switching on the input circuits permits the display of either channel A or channel B signals, in addition to both signals displayed at the same time. Another function allows the addition or subtraction of the two input signals. Let us explain how these functions are used.

Phase shift. Earlier in this chapter the discussion covered the phase-shift characteristic of audio amplifiers. When one uses a single-trace scope the probe is moved from the input to the output of the circuit being tested. The technician has to note or remember the exact values and waveforms being measured. An easier way to do this is shown in Figure 8-2. A dual-trace scope is used and permits observation and measurement of both the input and output waveforms at the same time. One test probe is connected to the input circuit of the stage being tested. The second channel test probe is connected to the output of the stage being tested. The two vertical attenuators are adjusted so that both signals are displayed. These two signals are compared for amplitude and phase inversion. This sytem is much easier than attempting to use a single-trace scope. Vertical positioning controls for each signal permit the technician to have the two displays shown on the same horizontal axis. They can then be closely compared.

Digital analysis. A dual-trace scope is very handy for the analysis of digital circuits. First, the shape of the wave used in digital circuits is ob-

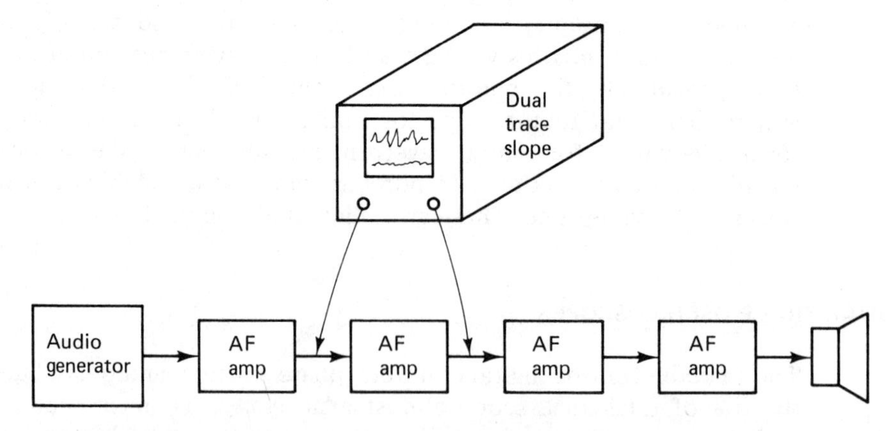

Figure 8-2 A dual-trace oscilloscope will display both input and output waveforms at the same time, for comparison purposes.

served. It should be a true square wave, such as the one shown in Figure 8-3. Both the leading and trailing edges of the wave must be free of any distortion, in addition to being vertical. The upper and lower horizontal portions of the wave must be straight and without any spikes or deviations.

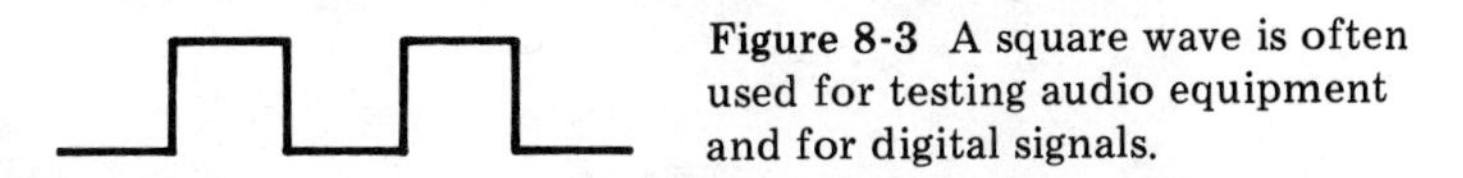

Figure 8-3 A square wave is often used for testing audio equipment and for digital signals.

The waveforms observed in a divider stage are shown in Figure 8-4. The upper waveform is the divided, or higher-frequency-rate signal. The amplitude, phase, and shape must be compared in these circuits. These divider circuit waveforms can be shown on a single-trace scope. It is much easier and far more accurate to compare them on the dual-trace scope. As more circuits in radios, turntable drives, and amplifiers are using digital techniques, the wise technician learns how to utilize this valuable piece of test equipment. The end result is faster analysis and better repairs.

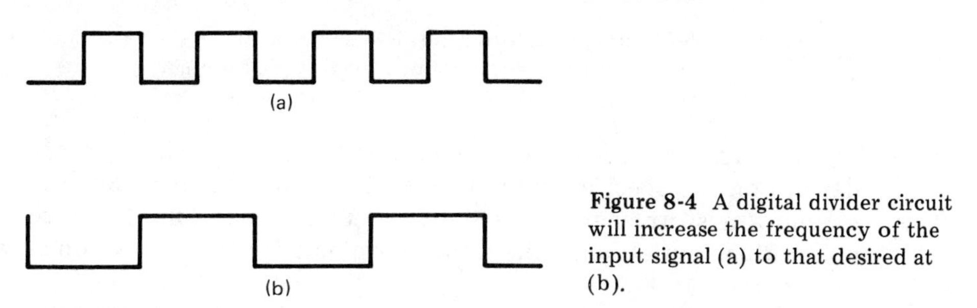

Figure 8-4 A digital divider circuit will increase the frequency of the input signal (a) to that desired at (b).

Power supply. A variable dc power source capable of producing from 1.0 to 20 V or so is desirable. Its current capacity should be at least 1.0 A. If one desires to repair car stereo equipment, a 13- to 14-V dc supply capable of providing at least 8 to 10 A is a necessity. The exact value of current available from the power supply should exceed the current drain requirements of the units being serviced.

The exact type of equipment to be serviced will determine the requirements for the bench power supply. Many of the higher-power output amplifiers on the market use a 45-V power supply. If these units are to be serviced, a suitable power supply is required. Here, as with car stereo units, the current capacity for the amplifiers must be considered. One may also need a dual-polarity power source for some amplifiers. It might be wise to have more than one power supply available. Each power supply could then be purchased or constructed to meet the needs of the system being serviced.

Variable autotransformer. A variable autotransformer is a requirement when servicing solid-state direct-coupled output stages. Variable autotransformers are available with a variety of voltage and current ratings. Almost all

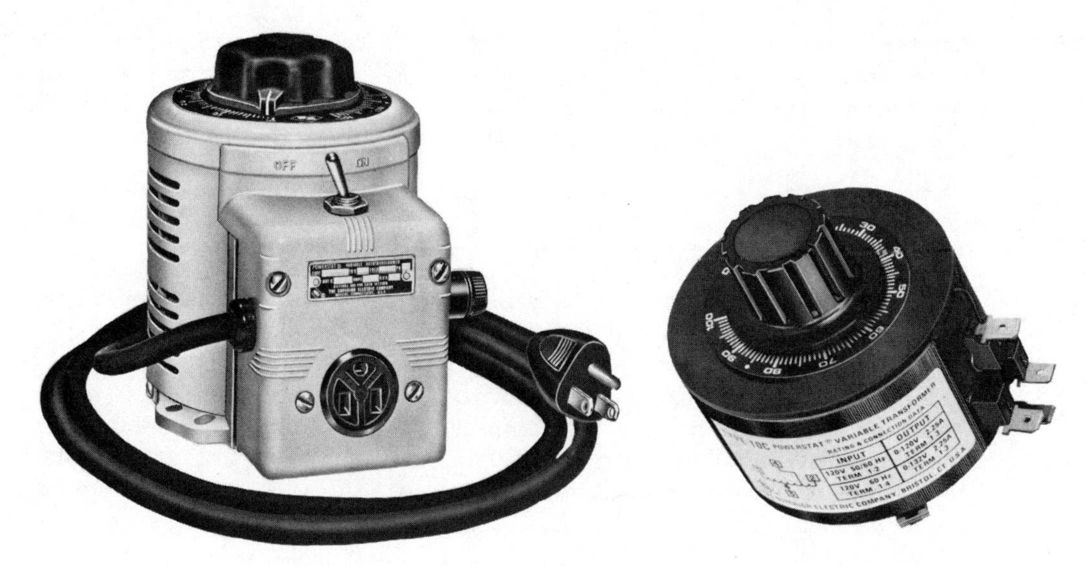

Figure 8-5 Variable autotransformers are used to control the level of the ac input voltage. (Courtesy Superior Electric Company.)

of those used in a service shop are rated at 120 V. The current rating will depend on the specific requirements of the equipment being serviced. Typical autotransformer units are shown in Figure 8-5. Some of these are available with ac outlets and others have terminals for the connection of wires.

Frequency counters. Frequency counters are becoming a common piece of test equipment. Historically, the frequency counter was found only in laboratories or installations that required accurate frequency measurement. The introduction of the integrated circuit in consumer products makes the need for a frequency counter a reality for service work. Oscillator, timing, and counter circuits need to be tested for frequency accuracy. A service type of frequency counter, similar to the one illustrated in Figure 8-6, is required for the circuits used in current production products. The frequency range of this type of counter should be up to about 100 MHz.

Figure 8-6 A frequency counter is becoming a very necessary piece of test equipment. (Courtesy Leader Instrument Corporation.)

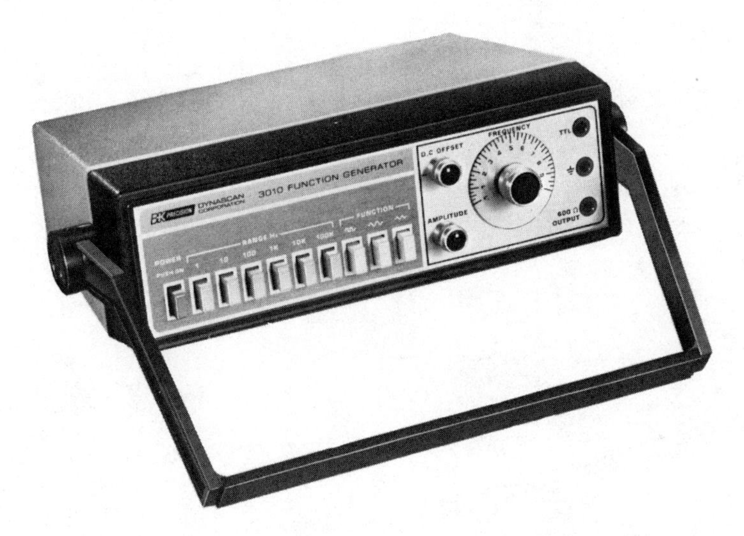

Figure 8-7 Sine, square, and triangular waves or functions. A generator is used to produce audio test signals. (Courtesy Dynascan Corporation.)

Signal generators. There are different schools of thought on the types of signal generators required for service work. A prime consideration is the identification of the frequencies to be created. Different signal frequencies are used for AM, FM, FM stereo, and tape players. The selection of specific test equipment depends on which of these signals are required. Signal generators are categorized generally as audio-frequency or radio-frequency generators. In addition, the audio generator is often called a *function generator.* One of these is shown in Figure 8-7. This generator is capable of producing sine, square, or triangular waveforms. The frequencies for these run from about 10 Hz to over 1 MHz.

A radio-frequency generator will produce sine waves in the radio-frequency range. This range is usually from 100 kHz to about 30 MHz. This generator will also produce a modulation signal. The combination of the modulation and the RF signal develops into an amplitude-modulated carrier.

Separate signal generators are available for FM signals. One generator that is able to produce all the signals required for FM and FM stereo is shown in Figure 8-8. This unit will create a carrier, with or without modulation, as well as the 19-kHz pilot signal and the L - R stereo component.

Some test equipment manufacturers package several individual units in one cabinet. The unit shown in Figure 8-9 is an example of this. This tester has both an AM and a FM signal generator. In addition, both carriers may be modulated with a 400-Hz tone. It also has a FM stereo generator that is able to create all the required FM stereo signals. Another nice feature is the inclusion of output meters. Two load resistors are included in the unit. These resistors act as a speaker load. It is possible to measure the power

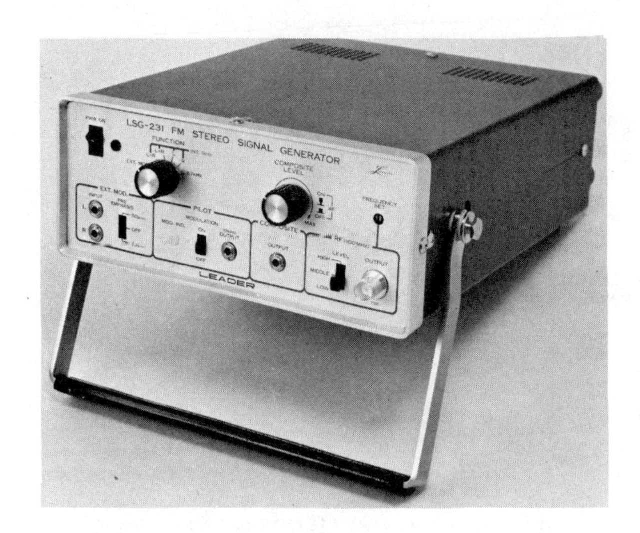

Figure 8-8 An FM stereo generator will produce constant valid test signals for FM tuners or receivers. (Courtesy Leader Instrument Corporation.)

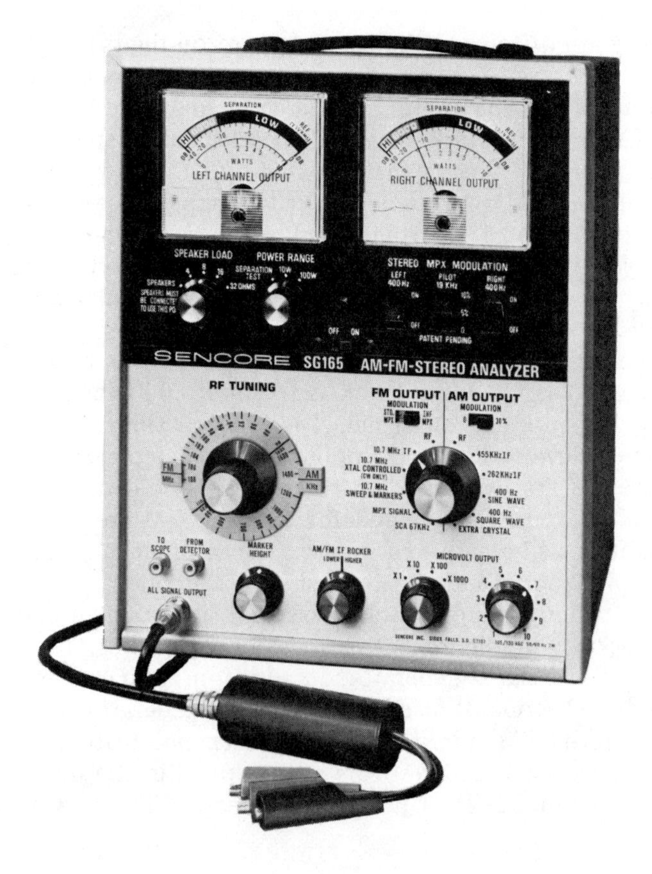

Figure 8-9 A composite AM–FM stereo signal generator will provide all necessary test signals for stereo servicing. (Courtesy Sencore, Inc.)

output of an amplifier when the meter and load leads are connected to the output terminals of the amplifier.

The service bench may have other pieces of test equipment. The units described represent a basic complement of equipment. Each service technician has his or her own set of ideas as to a specific brand of equipment to purchase. In addition, one may wish to build some of these from kits that are available. The more daring individual may even wish to develop some of these items from "scratch." Any of the above are acceptable as long as the units work correctly and produce a valid signal. The next thing to look at is how best to use this equipment.

BASIC TROUBLESHOOTING

There are different ways of approaching a troubleshooting problem. One must always keep in mind that the set did work at one time. One must also go on the basis that the set has not been redesigned by someone attempting to repair it. The most important factor in any sequence of repair is to think. The problem area should be identified very early in the troubleshooting process. Chapter 7 described this process. The technician is not supposed to do this work without any help. One of the best aids is the schematic diagram. This diagram will help identify blocks. It provides specific information about the circuit wiring in each block of the receiver.

Once the area of trouble is located, the technician has to know how to locate the specific component that has failed. When a modular receiver is being repaired the malfunctioning module must be identified. The process of identifying a malfunctioning module or circuit requires the use of test equipment. Selection and application of proper test equipment speeds up the repair process. Two basic methods of localizing a problem area are signal injection and signal tracing.

SIGNAL TRACING

Signal tracing requires the use of two pieces of test equipment. One of these is a signal generator. The other is an oscilloscope. The signal generator is used to develop a constant signal. The specific type of signal used will depend on the circuit that requires tracing. Stereo servicing limits the selection of this to either an audio or an RF signal generator. A connection for this setup is shown in Figure 8-10. An RF signal generator is used to establish a proper input signal. The frequency and amplitude, as well as the type of modulation, are obtained from the service literature for the specific unit.

Logic should be applied at this time. Once the signal generator is connected and both units are turned on, the technician analyzes the results of

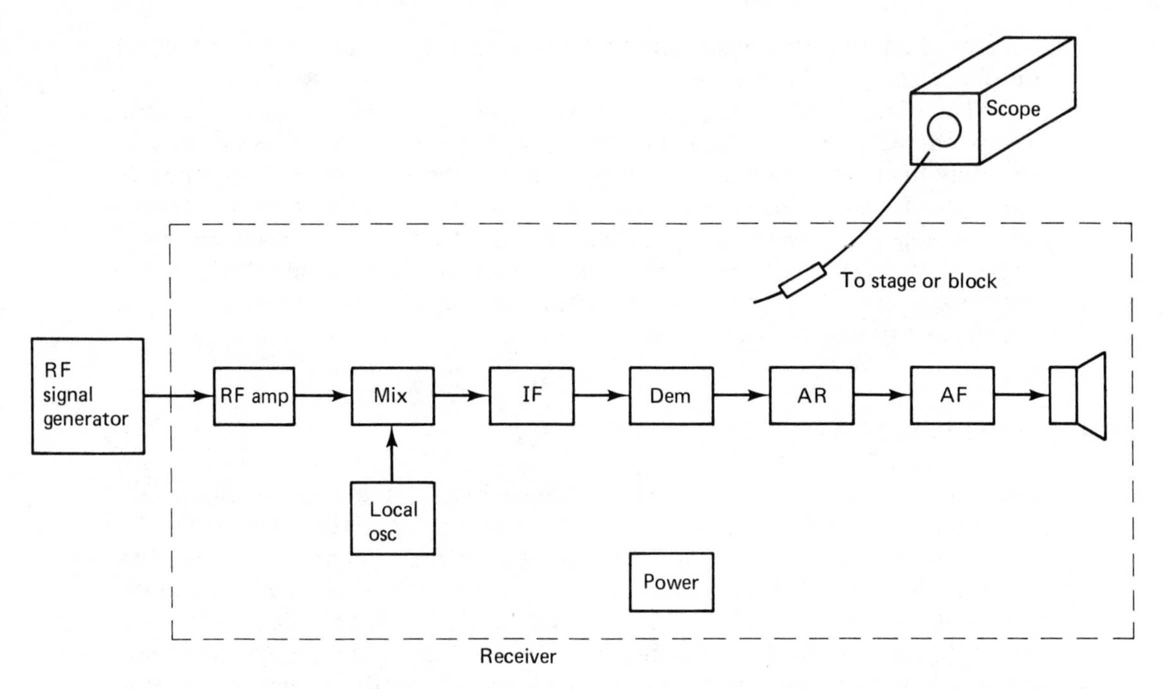

Figure 8-10 Signal tracing is done by injecting a known signal at the input and using a scope to trace it through the system.

the hookup. One listens for any sounds from the output. A low level of 60- or 120-Hz hum indicates that the power supply is probably functioning. Tuning the generator slightly off the generated frequency will help determine if the dial is incorrect and the unit functions properly. Another test is to tune the generator to the IF frequency. This step bypasses the local oscillator and helps to determine if it is working. In other words, the technician attempts to eliminate all the working blocks early in the procedure of troubleshooting. It is a waste of time and energy to test the working sections in order to locate a nonworking section.

A rapid method of troubleshooting a radio receiver is desired. This can be done by applying the rules for signal path troubleshooting. These rules are described in Chapter 7. The radio receiver can be considered a basic linear flow system with some exceptions. Approach it first as a linear system and then apply the rules for meeting or feedback paths if necessary. The rule for a linear path system is to make the first test at or near the middle of the system. The initial test for the receiver could easily be made at the output of the demodulator. Use an oscilloscope to make this measurement. If an audio signal is seen, all blocks from the antenna to this point are functioning. The problem area is considered to be between the test point and the output point. It, too, can be split in half by testing at its midpoint. This process is continued until only one block is suspected. Then other tests and measurements are made to locate a specific component that is malfunctioning.

This method of troubleshooting is rapid and eliminates wasted effort. The temptation to start the signal tracing after the first block instead of at the middle of the path is a great one. This is usually a waste of effort and time. If the problem happens to be in the RF amplifier, then the technician is very lucky. Good troubleshooting techniques are not a matter of good luck. They are a matter of developing excellent work habits and using these habits each time. Certainly, there will be some wasted effort. But overall, this method will produce accurate results and make the technician more efficient.

A trap to avoid in troubleshooting is the assumption trap. Do not assume anything. Make several different tests in order to prove that the circuit is working properly. Once you start to assume that sections are working, you will find yourself falling into a habit of wasting effort. Usually, one assumes that a section is functioning and then spends much time and energy making further tests of other sections only to return eventually to the assumed good area to locate the fault. Develop a logical, efficient procedure for diagnosis and testing. Follow this procedure and attempt always to improve on it. Share your techniques with others so that they, too, will be productive.

SIGNAL INJECTION

A second method of approaching the diagnosis of a defective electronic unit is called signal injection. This method differs from signal tracing in that a signal source is injected at each stage. The process starts just as any other logical troubleshooting approach. A linear system is still split at or near the midpoint. The difference lies in the manner in which the test equipment is used. When signal tracing, a signal is injected at the input to the system. It is then traced through the system using an oscilloscope or other signal-tracing device. When one uses the signal injection method, the test signal is injected at the point to be tested. This procedure is illustrated in Figure 8-11.

The only piece of test equipment required for this process is the signal generator. The generator must be capable of creating the correct RF, IF, and audio frequencies. It is adjusted to produce the desired signal. The test signal is then injected at the proper test point in the unit. When testing a radio receiver the most logical point at which to start is the output of the demodulator. If the proper audio signal is heard at the output of the receiver, all sections located between the test point and the output are working correctly. The generator is then adjusted so that a correct signal is injected at the output of the mixer. The procedure is to move from the first test point toward

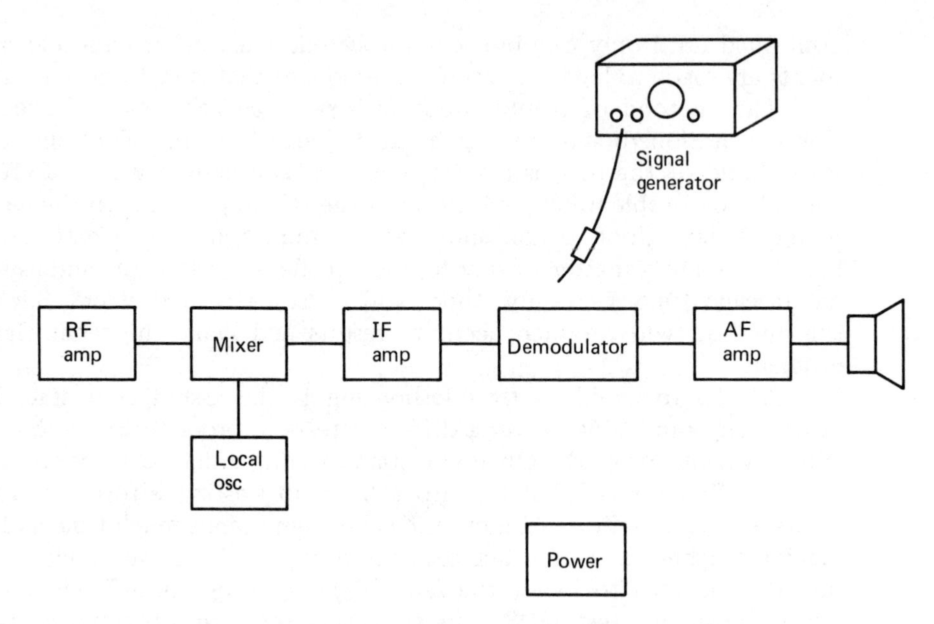

Figure 8-11 Signal injection uses a signal generator and its output is moved to each test point.

the input to the receiver. When the signal is lost, the problem area is between the current test point and the previous test point.

The approach used for signal tracing and signal injecting is similar in procedure. The basic similarity is the point at which the troubleshooting starts. The difference is the manner in which it continues. Tracing moves from the input toward the output circuits. Injecting moves from the output back toward the input of the block. Both methods are correct and proper to use. The advantage of using the oscilloscope is that waveform shape and amplitude are checked against manufacturers' service data. This cannot be done when signal injection procedures are followed. Procedures for either of these techniques are discussed thoroughly in the manuals provided by test equipment manufacturers. These manuals contain a wealth of valuable information. They should be required reading for all users of test equipment.

LOCALIZING THE DEFECTIVE COMPONENT

Procedures described earlier in this chapter may be applied to a specific stage. There are two processes that happen at the same time in a working electronic circuit. One of these is signal processing. The second is dc operation and current flow. Analysis of a defective stage requires analysis of both these processes. The circuit in Figure 8-12 is a good example of the kind of analysis done by a technician. Much of this thinking is done at a subcon-

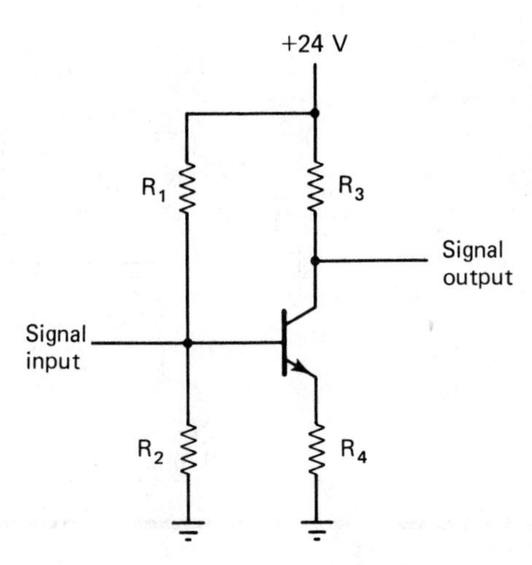

Figure 8-12 Common-emitter amplifier circuit used for analysis in the text.

scious level as skill in troubleshooting is gained. The circuit containing R_3, R_4, and $Q_{1E\text{-}C}$ is a series circuit. A source of 24 V will cause an electron current flow. Total current is unknown. As the current flows, voltage drops develop across the three resistances (R_3, R_4, and Q_1). The probable voltage at the collector is one-half of the source voltage, or about 12 V. This last statement is based on previous experience. The circuit is a basic common-emitter amplifier. A small signal at the base will produce a larger signal at the collector. The collector, or output, signal is inverted in this type of circuit.

The input circuit bias is established by R_1 and R_2. This voltage divider will set the operating point for the amplifier circuit. The voltage established at the base of the transistor depends on the exact values of these two bias resistors. This can be determined from the data found in the service literature for the unit. Another factor that is almost always constant is the voltage drop developed between the base and the emitter of a silicon transistor. This voltage is about 0.7 V. Measuring it when power is applied to the circuit should produce this value. Any voltage that is much higher than this value indicates a leaky transistor. A zero voltage value indicates a shorted transistor.

A second circuit to be analyzed is shown in Figure 8-13. This is a partial schematic of a more complex circuit. It contains a transistor, a crystal filter, and an IC. The specific information about the IC is not given in the service literature. Information about the function is given. The name of this IC describes its function. It is called an IF amplifier. The circuit is part of a FM tuner. One can assume that it will process and amplify an IF signal.

Many IC circuits use the term "gain" amplifier rather than either "voltage" or "current" amplifier. This is really a better term. The definition of an amplifier is that it is a device that uses a small amount of power to control a larger amount of power. In many instances the gain in the circuit is in the form of a current gain. This cannot be measured with ordinary test equipment. We are used to observing voltage gain in an amplifier on the

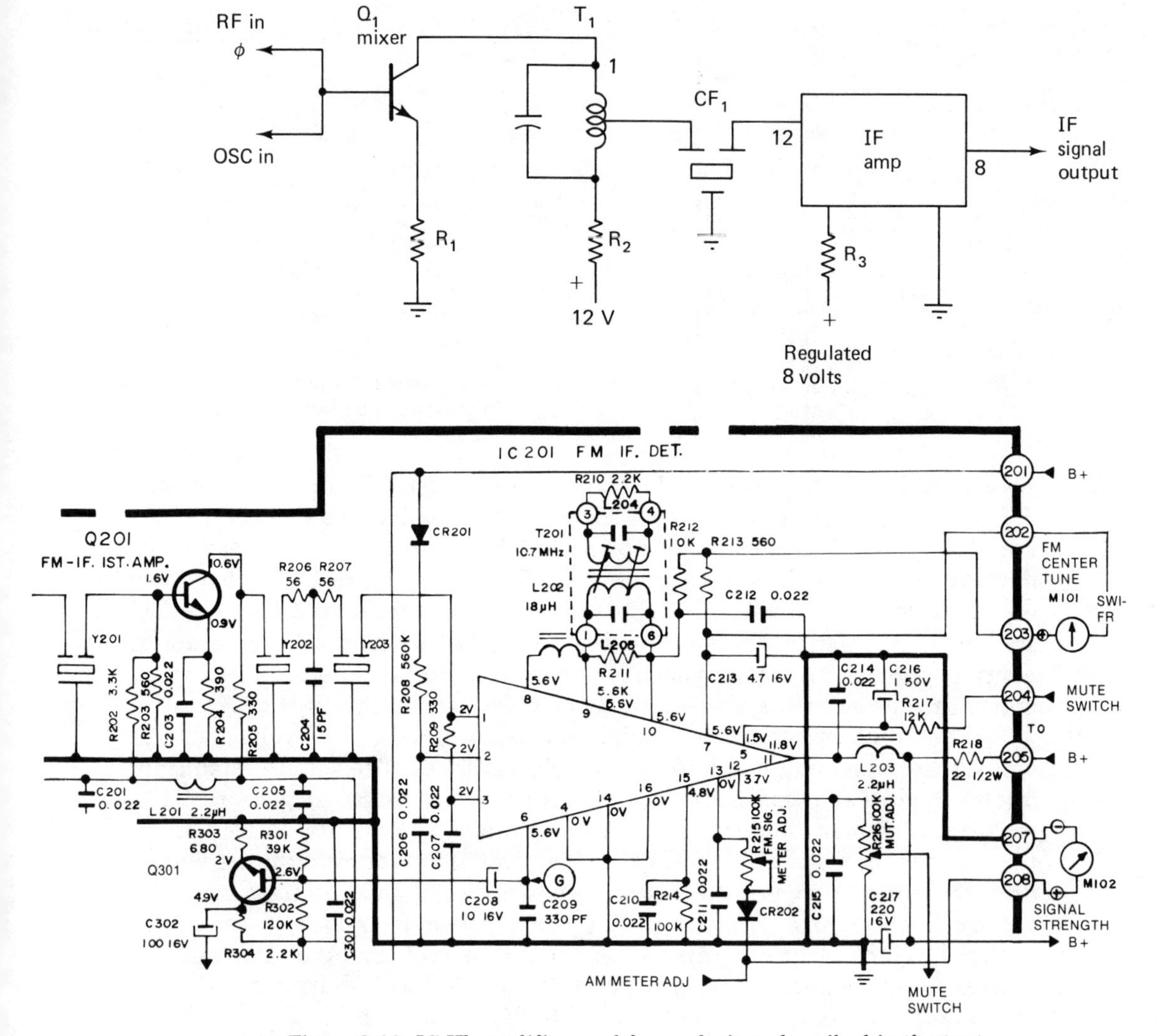

Figure 8-13 IC IF amplifier used for analysis as described in the text.

oscilloscope. A current gain is not an observable factor. The technician may falsely assume that a stage is malfunctioning if a voltage gain is not observed. Knowledge of circuit operation is important to analyze properly the information obtained from the receiver.

This circuit may be analyzed in two different ways. Both are correct and required. The easiest way to start is to consider signal processing. For most AM and FM tuners, signal information is *not* included on the schematic diagram. This means that the service technician has to know what these waveforms look like and also be able to guess as to the amount of amplification that occurs in the stage. Transistor Q_1 is a common-emitter amplifier.

This means that it should exhibit a voltage gain of some value. Because it is used as a mixer, it will not have a great amount of voltage gain. The FM signal does not readily exhibit the characteristics of phase inversion. This has to be assumed because the signal is symmetrical at this point. Checking with an oscilloscope should verify this assumption. One should expect to observe a slight voltage drop across the crystal filter CF_1. No information other than the name of the IC is available. The measurement of the signal input and output of the IC should show a voltage gain of a fairly large value. If this is true, there is no need to check this circuit further.

If the expected values are not found when the measurements are taken, one has to shift to current flow analysis to locate a defective component in the circuit. This is called *dc circuit analysis*. There are two separate dc circuits shown in this schematic. One of these provides power to the mixer transistor Q_1. The second provides the proper regulated voltage for the operation of the IC.

The dc analysis of the mixer circuit should start with a voltage measurement at the collector of the transistor. There is relatively little dc resistance in the IF transformer T_1. The voltage measured at the collector of this transistor should be just under the supply voltage value. The voltage at the emitter of the transistor should be close to zero. It will often be about 1.0 V. If either of these values is incorrect, resistance checks should be made to determine which component is defective.

The analysis of the IC circuit is similar. First, the source voltage is measured. Next, other operating voltage values are measured and compared to those values given in the service literature. Any major differences indicate a defective component. Specific circuit analysis will then identify the component that has failed.

Probably by this time you are wondering if you will ever become so expert as to be able to think about a service problem in the manner described above. Do not worry about doing this. The process is one that develops. The idea is to start to develop good work habits. As your experience increases, your logical troubleshooting analysis will also become better. Try to review a troubleshooting procedure after the problem is corrected. See if you can develop a better way of solving the problem for the next time. It might help if you discussed the problem and your approach to the solution with other technicians. After all, why do you have to rediscover that which others are using?

TECHNICAL INFORMATION

Information required for the rapid repair of a receiver may be obtained from several sources. There are two major sources for this information. One of these is the receiver manufacturer. Almost all of the major manufacturers have supply outlets. These are usually located in, or near, major cities. A

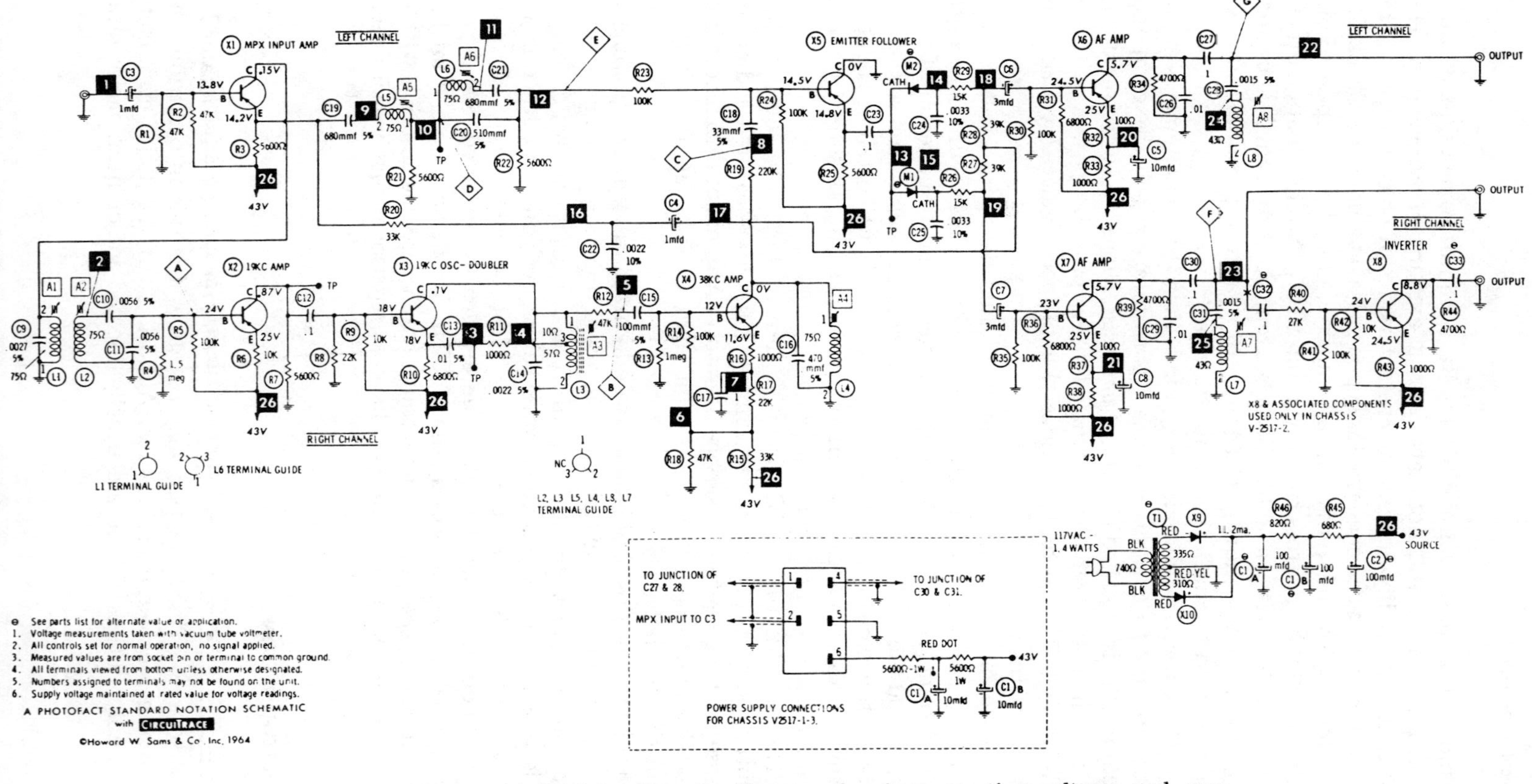

Figure 8-14 Partial schematic diagram showing operating voltages and component value information. This information makes the servicing of the unit convenient.

phone call to one of these outlets will establish the availability and cost of this information.

A second major source is information published by the Howard Sams organization. This company publishes a series of packets of technical information. They are called "Photofacts."* Each packet contains schematic, alignment, and replacement parts information for a group of consumer electronic devices. This material is available from many local independent electronic parts distributors. Many libraries also have this information available.

The information included in the service literature available from either source includes waveforms, operating voltage values, and component information. A section of one of the schematic diagrams is shown in Figure 8-14. A full schematic is too large to reproduce in this book and still be readable.

Do not attempt to repair any electronic unit unless you have the service data available. The extra time spent in obtaining this valuable information pays off in time saved when making the repair. Each manufacturer has its own way of using the components in the circuit. Parts layout varies greatly from one set to another. This is true even for sets made by one manufacturer. Analysis time is not wasted time. It is actually time well spent. This time will ultimately earn money for you as you develop good work habits.

QUESTIONS

8-1. Name the basic test equipment used for stereo servicing.

8-2. Why should a dual-trace oscilloscope be used?

8-3. When is a variable autotransformer used?

8-4. What circuits require the use of a frequency counter?

8-5. Explain how signal tracing is done.

8-6. Explain how signal injection is done.

8-7. Name some sources for technical service literature.

8-8. What is the proper dc voltage measured between the base and emitter of a PNP transistor? for a NPN transistor?

8-9. Dc analysis shows source voltage at the collector of a common-emitter amplifier. What parts are suspected?

8-10. Dc analysis shows zero volts at the collector of a common-emitter amplifier. What parts are suspected?

*Trade name for H. W. Sams publication.

Chapter 9

Power Supplies

Probably the most important section of every electronic device is its power supply. The power supply must be able to provide the correct amount of power for proper operation of the device. The power supply must also be able to deliver the correct voltage and current values to the load circuits. In addition, the power supply must deliver voltages of the proper polarity and do all of these things with little or no fluctuation in voltage values.

There are several circuits that are in common use in power supplies. Most of them operate on similar principles. A block diagram for a basic power supply system is shown in Figure 9-1. A voltage step-down transformer is used in order to reduce the supply voltage of 120 V to the level required by the system. Most solid-state devices operate on voltages of less than 50 V. The transformer is connected to a rectifier block. Its purpose is to change the alternating source voltage into a direct-current voltage. The output of the rectifier is a pulsating dc voltage. The pulsations need to be smoothed to obtain the dc values required by the load. This is accomplished by use of a filter system. Once the dc has left the filter, it is distributed to the proper blocks in the device. Some circuits require very close control of the output voltage. When this is a requirement, a voltage regulator block is used. More than one voltage regulator may be required for a specific device. This is true when different regulated voltage values are required. One last output from the system may be connected directly from the power transformer. This is the section of the power supply used to light the filaments of any vacuum tubes. An ac source is usually used for this purpose.

The basic system just described operates from a 60-Hz power source.

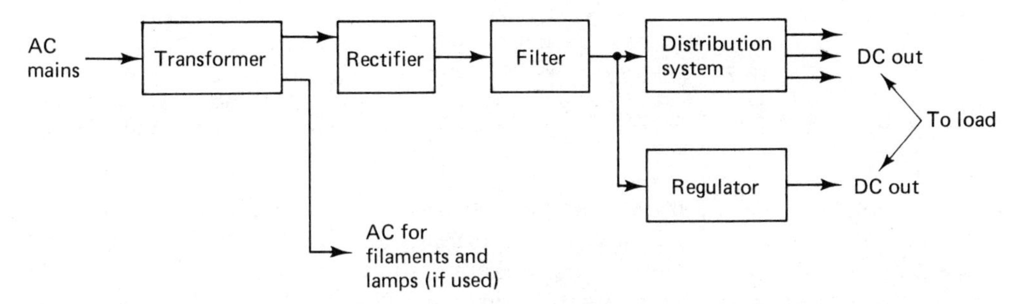

Figure 9-1 Block diagram for a power supply system.

Components required to develop high power levels are large, heavy, and costly. Technological developments in recent years have led to the introduction of a different type of power supply. It is called a switching type of power supply. The basic system uses a high-frequency oscillator and some high-speed transistors to develop the correct operational voltages. Details as to how this system functions, the operation of the basic power supply circuits, and how to troubleshoot these systems are discussed in this chapter.

THE TRANSFORMER

Almost all nonportable electronic devices are transformer operated. The transformer changes the input voltage to the desired value and also serves to isolate the device from the voltage source. Transformers are designed to meet the specific requirements of the load to be operated. Both voltage values and current values have to be considered when a transformer is selected. The use of an underated transformer will lead to overheating and eventual failure of the transformer. The ratio of turns of wire used for the primary and secondary windings determines the voltages available at the secondary. A turns ratio of 1:2 will produce a secondary voltage that is twice the value of the primary voltage. The size of the wire used for the windings determines the current capacity of the transformer.

It is often desirable to have more than one secondary winding on the transformer. This is not a problem. The manufacturer uses two separate sets of windings for this purpose. Each winding has its own set of electrical specifications. The electrical ratings of two secondary windings are used to determine total power characteristics for the transformer. A schematic diagram for a dual secondary step-down power transformer is shown in Figure 9-2. The primary is operated at an input of 120 V. The current consumption is 0.5 A. Power consumption for the primary is 60 W. The secondary windings produce 12 and 24 V. Current ratings for secondary 1 is 3 A. The power consumption is 36 W. The other secondary winding produces 24 V at 1 A. Its power consumption is 24 W. The power consump-

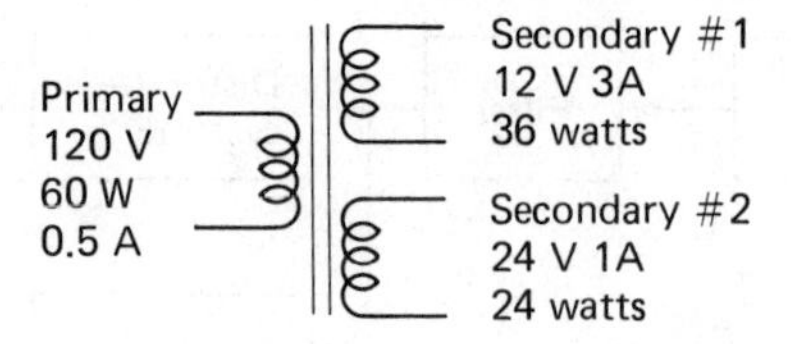

Figure 9-2 Schematic diagram for a dual secondary voltage step-down transformer.

tion for both secondary windings is 60 W, the sum of each winding's power. This value is also the rating of the primary. In this example an efficiency rating of 100% is assumed. In a practical circuit the efficiency is less than 100%; therefore, the rating for the primary will exceed 60 W.

One major application for a transformer is its isolation properties. There is a definite need to separate electrically the power and the circuits being tested. Many pieces of test equipment have one side of their ac power line connected to circuit common. The "common" test level is also connected to this point. There are many instances where one side of the power line on a piece of electronic equipment is also connected to circuit common. It is possible to connect the plug on the power cord so that either lead of the line cord is connected to circuit common. When the condition shown in Figure 9-3 occurs, the two common leads are really *not* connected to the same side of the electrical source. A difference of 120 V exists between them. The cases or chassis of the two units poses a death trap to the unsuspecting technician. If the technician touches both units at the same time,

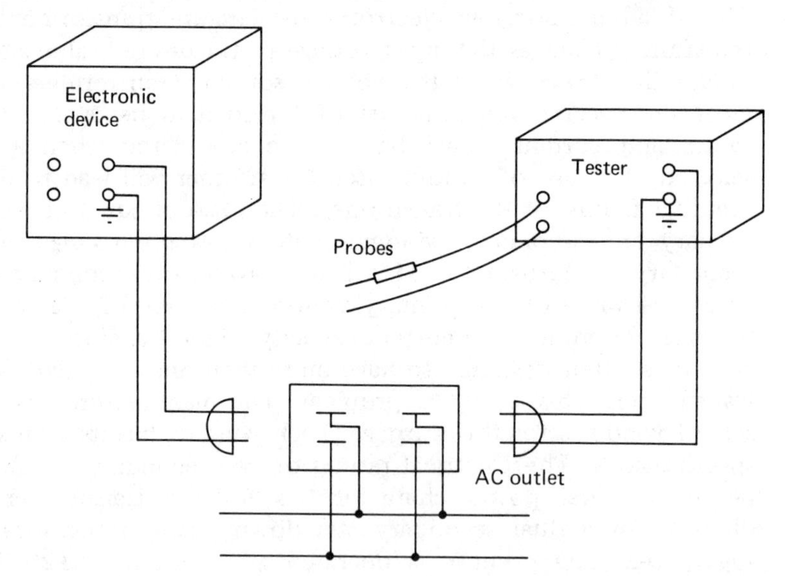

Figure 9-3 Common or ground connections must be made correctly. This shows them being made to opposite sides of the power line. This can cause shock or death and result in equipment damage if an isolation transformer is not used.

a current path is created through the body. An electrical current of 0.010 A (10 mA) can cause death under certain conditions! The wise technician does not allow this condition to exist. The use of an isolation transformer that is connected between the power source and the electronic device to be tested eliminates this problem. The use of an isolation transformer should be standard practice for testing all electronic devices that are powered from the ac source.

If the conditions just described existed but the technician was fortunate enough not to touch both chassis at the same time, another potential cause of trouble exists. When the two chassis are connected together by the common lead from the tester, a "ground loop" exists. Current flows because the two commons are not at the same electrical potential. The result of this ground loop is often damage to either the tester or the unit being tested. This adds to the cost of the repair or disables required test equipment. Neither situation is desirable.

RECTIFIER SYSTEM

There are several rectifier systems in common use today. These include half-wave, full-wave, full-wave bridge, and voltage-doubling types of systems. Each of these is used in some stereo units. Each system has characteristics that are found only in the one system. Let us examine each of the rectifier systems.

One of the things that appears to confuse people when they study rectifier systems is the difference between input and output voltage values. These values are really not different. The reason for the apparent variation is that two different systems for measuring the same voltage are used. Common acceptance of voltage levels is the use of *rms* values. The output of the rectifier system uses *peak* values to describe its voltage level. To change from one system to the other, the rms value is multiplied by 1.414. An rms value of 24 V is the same as a peak value of almost 34 V. This difference appears to provide an increase in voltage. In fact, it is the same value, but a different set of terms is used to describe it.

The description of rectifier circuits in this chapter uses the rms value to describe the ac input voltage. Output voltages, being dc, are described as peak values. Keep this in mind as you read the balance of this chapter.

HALF WAVE SYSTEM

A schematic for a half-wave rectifier system is shown in Figure 9-4. It consists of one rectifier diode, a load, and a source. The output waveform is one-half of the input wave. The frequency of the output wave is 60 Hz.

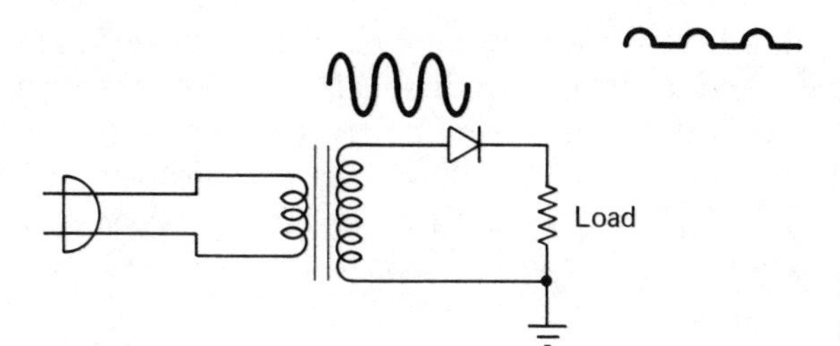

Figure 9-4 Schematic diagram for a half-wave rectifier system.

Current flow is possible only when the diode is forward biased. This occurs when the upper part of the transformer secondary is positive with respect to circuit common. During that portion of the ac cycle the resistance of the diode is very low, conduction occurs, and a large voltage drop develops across the load. Current flows through the load and work is accomplished.

When the polarity of the ac wave at the secondary of the transformer is reversed, a different set of operating conditions exists. The diode is now reverse biased. It develops a very high internal resistance when in this condition. The resistance is many times higher than that of the load. Almost all of the voltage applied to the diode and load now develops across the diode. There is no voltage drop across the load and, as a result, no current flows. This means that the load is turned off during this portion of the operating cycle. Additional circuitry is required in order to have load current flowing during all of the duty cycle. This is accomplished by the addition of a filter capacitor in parallel with the load.

FILTER CAPACITORS

The term "filter capacitor" is perhaps a bit misleading if one does not know how it functions in the circuit. A capacitor is designed so that it will hold an electrical charge. The charge will be held for a reasonably long period of time. The capacitor will give up its charge when it is connected to a current path. The charging time and the discharging time for any capacitor depend on the amount of capacitance and the size of the resistance in the circuit. This time is called the *capacitor time constant factor*. The formula for this is $T = RC$, where T is the time in seconds, R is the resistance in ohms, and C is the capacitance in farads. This represents the time required to charge the capacitor to 62.5% of its full value. Five time constant units are necessary for a full charge. In other words, the larger the capacitor, the longer it takes to charge or to discharge. This factor is important in a power supply filter circuit.

The waveforms shown in Figure 9-5 will aid in the explanation of how the filter capacitor functions in the circuit. The upper waveform represents the output across the load without a filter capacitor in the circuit. The dashed lines in the middle waveform represent the charge and discharge of

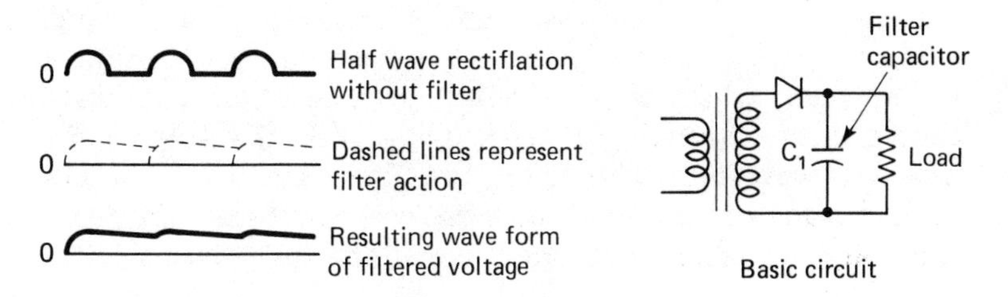

Figure 9-5 Use of a filter system reduces the voltage variations, or ripple, to a usable level.

the capacitor. The bottom waveform represents the dc output across the load as a result of diode and capacitor action.

The variations in dc output voltage are called *ripple voltage.* This value should not exceed 1.0 to 1.5% of the total voltage developed by the rectifier system. When a filter capacitor fails or is undersized, the ripple voltage increases. This is one test the technician makes as the electronic device is analyzed and repaired. Filter circuits used in electronic power supplies are often more complex than the one illustrated. The purpose of any filter system is to smooth out unwanted variations in voltage.

FULL-WAVE RECTIFIERS

The full-wave rectifier system is designed to operate on both halves of the ac cycle. The systems used can be classified as full-wave center tapped or as full-wave bridge rectifiers. Both systems are capable of producing a pulsating dc output voltage that has twice the frequency of the input waveform.

Full-wave center tapped. The schematic diagram for this system is shown in Figure 9-6. A transformer with a center-tapped secondary winding is required for this circuit. There are two rectifier diodes used in the system. Alternate halves of the transformer winding are used during each half-cycle of the input wave. Normally, the center-tapped connections of the transformer are considered as circuit common. Current flow in this circuit is dependent on the polarity of the ac voltage applied to the diodes. When one diode is forward biased, the other diode is reverse biased. Current flow is only through that half of the circuit that has the forward-biased diode. The

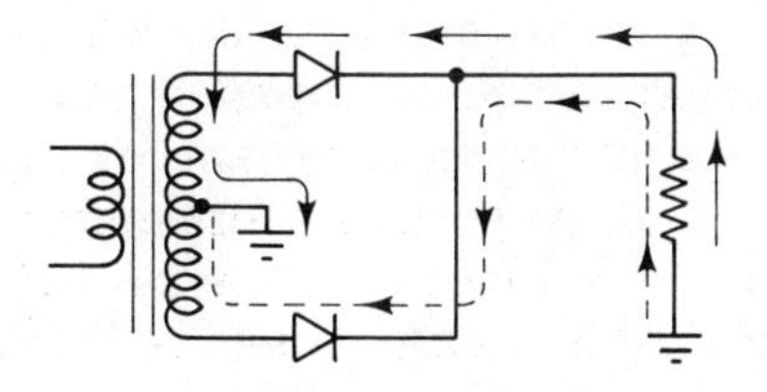

Figure 9-6 Schematic diagram for a full-wave system using a center-tapped transformer secondary.

current paths for each diode are as shown in this drawing. All the current flows through the load in the same direction. The output voltage is applied to the load, which is represented by a resistance. Its waveform appears as shown in the diagram. Filtering of the output voltage is required to make it into a dc waveform. The size of the filter capacitor for a full-wave circuit can be less than that required for a half-wave rectifier. The reason for this is that the charge and discharge time is less in the full-wave circuit.

Full-wave bridge. A full-wave bridge rectifier system is shown in Figure 9-7. This system uses four diodes. Two are used for rectification during each half of the ac wave cycle. The system does not use a center-tapped secondary on the transformer. The current paths for each half of the ac input wave cycle are shown in the drawing. The resistance shown represents the load. Current flows through it in the same direction, regardless of the polarity of the ac source voltage. The requirements for a filter capacitor are the same as those for a full-wave center-tapped rectifier system.

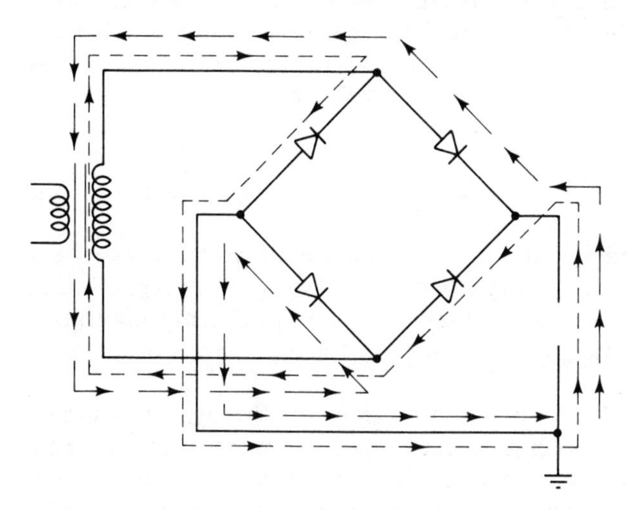

Figure 9-7 Schematic diagram for a full-wave bridge rectifier system using four diodes and no center-tapped transformer.

POWER SUPPLY POLARITY

The needs of the electronic load dictate the polarity of the power supply. Rectifier systems are designed so that their output voltage can be either positive or negative. The methods for doing this are illustrated in Figure 9-8. Part (a) of this drawing shows a system in which the positive output from the rectifier is connected to circuit common. This system is a reversal of the normal method of wiring a rectifier system. When a positive common is used, one has to be careful of the polarities for common between systems. It is possible to develop an overcurrent situation or destroy circuit components if two different polarity systems are connected together.

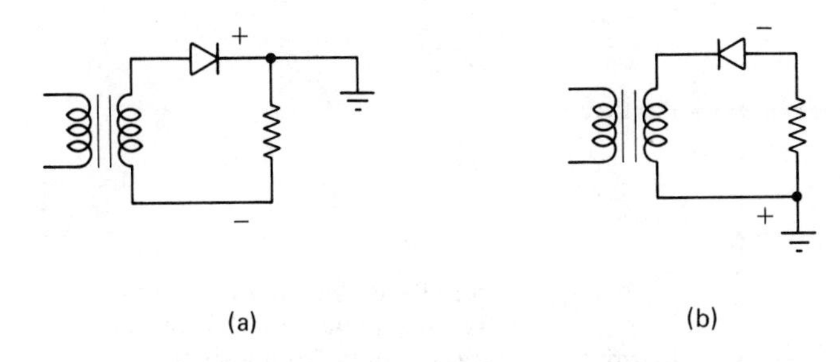

Figure 9-8 The polarity of the output of the power supply depends on the point of reference, or common, in the system.

The system shown in part (b) of the figure also develops a positive common. In this circuit the rectifier diode is reversed. Current flow is in reverse direction from its normal flow. When using either system one has to be aware of the polarities of the rectifier system and the filter capacitors. The filter capacitors are of particular concern. Electrolytic capacitors used in rectifier systems are very sensitive to voltage polarity. They are marked with polarity signs for each lead. The electrolyte in the capacitor will form a gas when connected in reverse to a power source. The gas expands and can cause an explosion as it tries to escape from the capacitor container. The end result can cause damage to the equipment and the technician.

DUAL POLARITY SOURCES

There are circuits that require both a positive and a negative voltage source. This can be done as shown in Figure 9-9. The system shown in part (a) illustrates one form of a dual-polarity supply that develops two output voltages which have the same amplitude but opposite polarities. Two diodes are connected to the same leads of the transformer. The diodes have opposite polarities. Each conducts during one-half of the ac wave cycle. Another method of doing this is shown in Figure 9-10. The circuit shown is a bridge rectifier circuit, but any non-center-tapped rectifier system can be used for this purpose. The circuit is unique in that there is no common connection from the transformer. The transformer leads "float" in this circuit. A capacitor and resistor voltage-dividing network is used across the dc output leads from the rectifier. The capacitors and the resistors must be of equal value. This is necessary to develop an equal voltage drop across them. The center

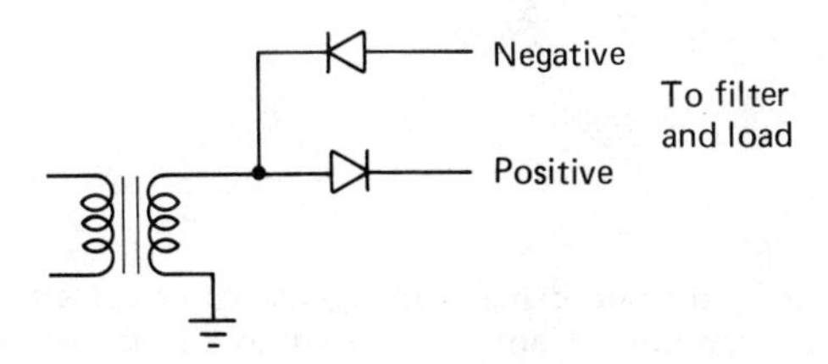

Figure 9-9 The possibility of obtaining dual-polarity values from a transformer depends entirely on the way it is wired.

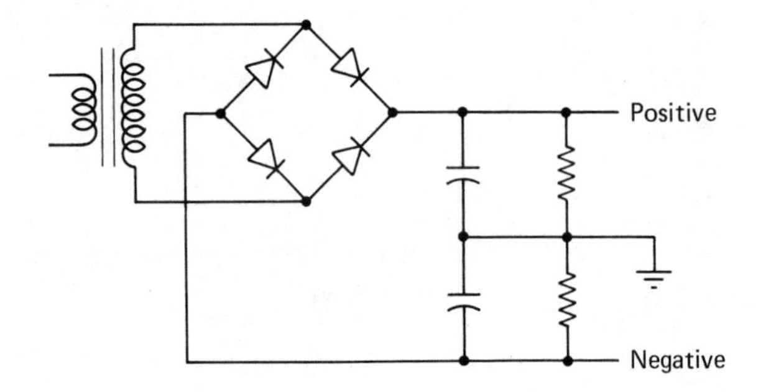

Figure 9-10 Another method of obtaining a dual-polarity and dual-voltage supply from a single source.

connection of the circuit is connected to circuit common. Both a positive voltage and a negative voltage are available from this type of circuit.

A circuit that is capable of producing a dual output voltage is shown in Figure 9-11. There are two similar circuits shown in this drawing. Both require a center-tapped secondary on the power transformer. Circuit (a) has both diodes connected, so that the polarity of the output voltage is the same. The value of the voltage obtained from the full secondary winding is twice that obtained from one-half of the winding.

One might feel that the voltage-doubler circuit is able to produce something for nothing. In this case the something is an output voltage that is twice that available from a rectifier circuit. The truth is that this assumption is wrong. Voltage-doubler circuits are able to make the output voltage twice that available from regular rectifier circuits. This is done by charging a capacitor during one-half of the ac cycle and using this charge to boost the dc voltage to a higher-than-normal output voltage. This system will work with both half-wave and full-wave rectifier circuits.

A schematic diagram for a half-wave voltage-doubler circuit is shown in Figure 9-12. This type of circuit is readily identified by the use of the series input capacitor C_1. This capacitor is charged by the applied voltage when point B is positive and point A is negative. Current flow will occur through diode D_2 and the capacitor will be charged. The capacitor plate at point A becomes negative during this half of the cycle. When the polarity of the applied voltage is reversed by the action of the generator, point A

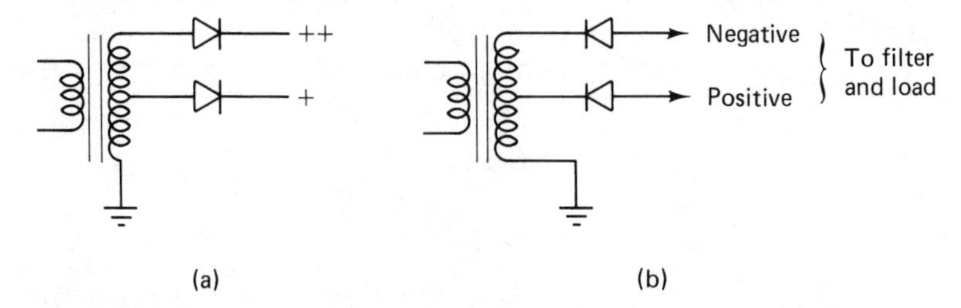

Figure 9-11 Dual-polarity output using a bridge rectifier system. The point used for circuit common is not connected to the transformer.

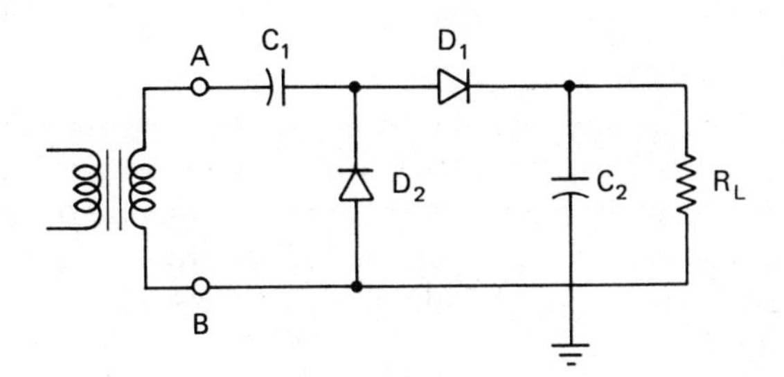

Figure 9-12 A half-way voltage-doubler system is dependent on the ability to charge capacitor C_1.

becomes positive and point B becomes negative. Current flow is from point B through the load and diode D_1 to the capacitor C_1. This capacitor has a charge that is equal to the applied voltage. This charge is now series connected to the ac input of the circuit. The net result is the addition of the charge voltage and the voltage applied to the circuit from the ac source or transformer. This doubles the applied voltage and also the dc output voltage. Capacitor C_2 is a filter capacitor in this circuit. Its purpose is to smooth out variations in the rectifier output voltage.

A schematic diagram for a full-wave voltage doubler is shown in Figure 9-13. This circuit requires three capacitors. C_1 and C_2 are series wired between circuit common and point A, the ac input to the circuit. The ac input to the doubler is connected across capacitor C_1. In general, the circuit does its doubling action by alternately charging capacitors C_1 and C_2. This is done during alternate half-cycles of the ac wave. Circuit action is as follows. When point B is negative, the current flow path is from that point through capacitor C_1, diode D_1, and back to point A. This charges C_1 to the value of the applied voltage. When the ac polarity reverses, point A becomes negative. Current flow is now through diode D_2 and C_2 is charged. These two capacitors are now able to provide a voltage charge that is equal to the applied voltage as additional cycles of the ac wave occur. The load is connected across the series-connected capacitors. The value of the voltage applied to the load is equal to the sum of the voltages developed across C_1 and C_2. In most circuits this is double the voltage available at the ac input to the circuit.

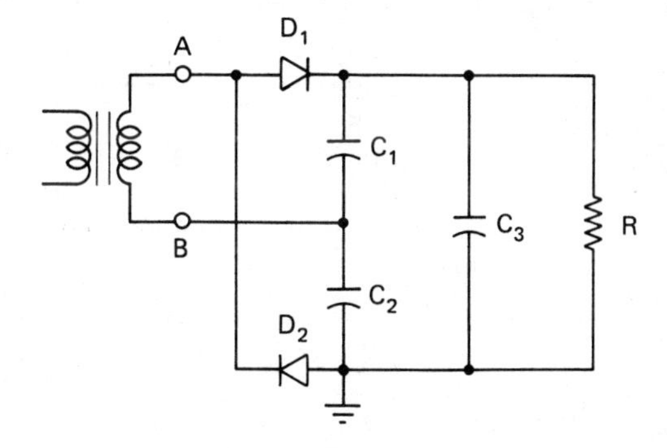

Figure 9-13 A full-wave voltage-doubler circuit requires the charging of the two capacitors, C_1 and C_2.

Efforts to reduce costs, weight, and size, as well as improving the efficiency of power supplies, have resulted in the introduction of a new type of power supply system. This is called a switching power supply. Someone might argue the point that all dc power supplies that are rectifiers are switching types. This is true, but standard half-wave and full-wave diode rectifier systems are not considered switching types.

The principal difference between the traditional power supply system and a switching power supply is the frequency of operation. A 60-Hz power supply operates only on that specific frequency. Components required for operation of the system at that frequency are large and usually quite heavy. Large amounts of copper wire and heavy laminations for the power transformer are basic requirements. The military services found that it was possible to provide adequate operational power when frequencies higher than 60 Hz were used. Many aircraft used generators that operated at a frequency of 400 Hz. The frequencies of operation were limited by the ability of the components to perform correctly.

The introduction of the transistor has allowed the development of high-frequency power circuits. A block diagram for one such circuit is shown in Figure 9-14. The input to this circuit is ac directly from the ac source. It is rectified and filtered to produce a dc operational voltage. This dc voltage is used to operate a high-frequency oscillator. A typical operational frequency for a system of this type is about 20 kHz. The output of the oscillator is then applied to a power transformer. A general rule about transformers is that their relative size will decrease as the operating frequency is increased. The design for a power transformer capable of operating at 20 kHz makes it only a fraction of the size and weight of its 60-Hz counterpart.

The secondary of the transformer is designed to provide voltages and currents for correct operation of the specific electronic device. In addition, the filter capacitors required for operation at 20 kHz are much smaller than those required for 60- or 120-Hz operation. Capacitance values for low-frequency systems range between 20 and 5000 μF. Filter capacitors for high-frequency systems are often less than 1.0 μF. The smaller capacitors

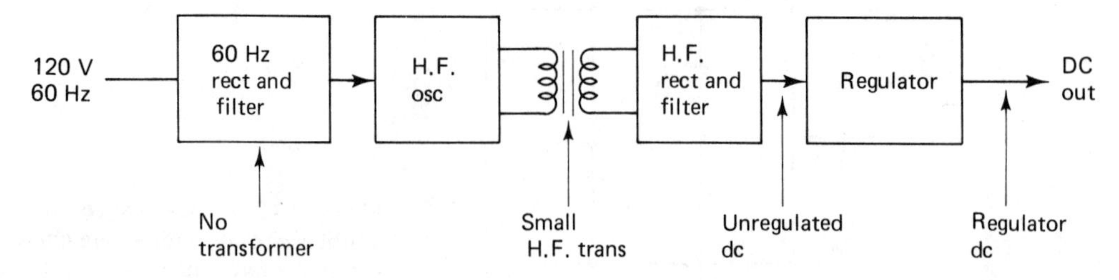

Figure 9-14 Block diagram for a high-frequency switching power supply.

used at the higher frequencies often reduce space requirements for the power supply to less than 30% of that required for lower-frequency operation.

REGULATION OF THE POWER SUPPLY

After the correct operating voltages and currents are created, they often are regulated. Regulation is necessary to maintain specific power levels in the system. A nonregulated supply will provide a higher than normal output voltage when there is no load connected to it. As the load requirements increase, the voltage tends to drop to lower values. As the load demand increases further and its equivalent resistance decreases, the voltage available from the nonregulated source decreases to even lower levels. A load resistance of less than 1 Ω will require large quantities of current and results in a low voltage.

Power supplies can be regulated. Regulation means that a constant voltage or current is maintained over a wide variety of load demands. Three types of regulators are in common use in the stereo unit. Two of these use discrete components. The other is available as an integrated circuit and all components are housed in one package. Let us examine all these types of regulators.

Zener regulation. One of the simplest forms of electronic regulators uses a zener diode. The schematic for this circuit is shown in Figure 9-15. A zener diode is able to act as a voltage regulator when it is reverse biased. There is an area of operation for the diode called the *zener region.* The diode will maintain a specific voltage level within the limits of its current capability. When placed in series with a resistor R_2, the zener diode forms a voltage-regulating circuit. The dc supply voltage must be higher than the desired regulated voltage. Both the zener diode and the series resistor must be able to handle the current and power in the circuit. The load is connected in parallel with the zener diode.

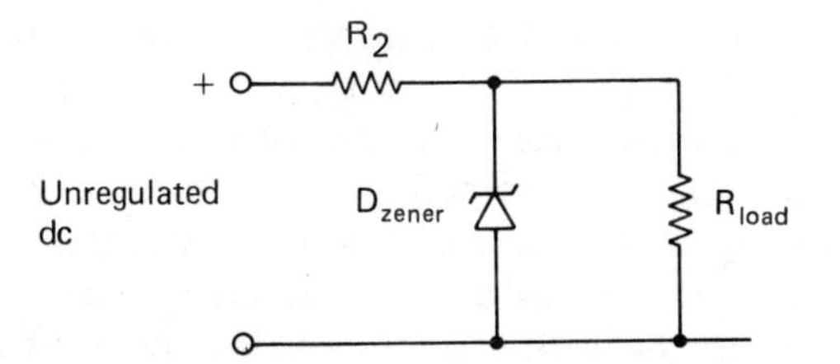

Figure 9-15 A zener diode is used as a voltage regulator in systems that do not have large changes in voltage or current.

Transistor regulators. A second discrete component circuit uses a transistor as a regulating device. This circuit is shown in Figure 9-16. A transistor is series connected between the load and the source of unregulated dc. The base of the transistor is held at a constant value by the use of the voltage

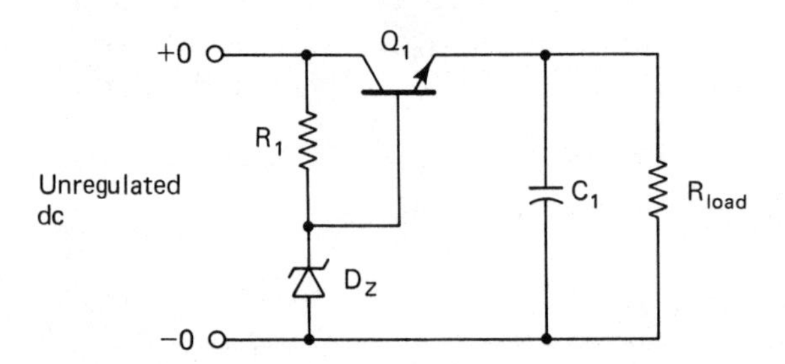

Figure 9-16 A transistor regulator circuit will offer better regulation over a variety of operating conditions.

divider network R_1 and D_z. When this type of circuit is used, the base-to-emitter voltage is held to a difference of about 0.7 V, with the emitter having the lowest voltage. If the zener regulator is designed to operate at 12.0 V, the emitter voltage will be 12.0 less 0.7, or 11.3 V. This circuit is designed to maintain a constant output voltage over a wide variation of current requirements. The current-carrying capacity of transistor Q_1 will act to limit total current flow. The power dissipation of the transistor is also a limiting factor. Heat develops in the transistor due to current flow and voltage drop across the emitter-to-collector connection. This heat has to be dissipated away from the transistor or it will be destroyed. A heat sink or a fan are often used to keep the transistor cool.

IC voltage regulators. A more sophisticated voltage regulator is now in use. This is the IC type of regulator. A schematic diagram and packages for this type of regulator are shown in Figure 9-17. This is a much more complex circuit, yet it occupies less space and usually costs less than its discrete component counterpart. The device has three connections. These are input, output, and common, or ground. Regulators like these are available with either positive and negative voltage output connections. They are available for a variety of output voltages. These voltages range from 5.0 to 24 V. Similar regulators have a variable output capability. In any case, the dc input voltage must be higher than the output voltage in order to have the desired regulation.

Switching regulators. Another recent addition to the regulator family is the switching type of regulator. One disadvantage of all other types of regulators is that there is a relatively large amount of heat developed during the regulation process. This is in effect wasted energy. The switching type of regulator tends to minimize the heating loss of the other regulators.

Conductive devices, such as transistors, will operate with less average heat dissipation when they can be switched on and off rather than have a constant operational level. The use of a pulse or square-wave generator to control operating conditions is the basis for less heat and more efficiency in the switching regulator. A schematic diagram for one type of switching regulator is shown in Figure 9-18. This circuit uses an error amplifier to

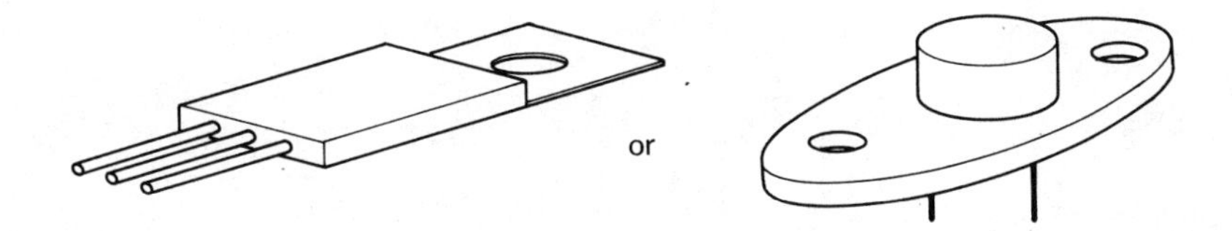

Equivalent schematic.

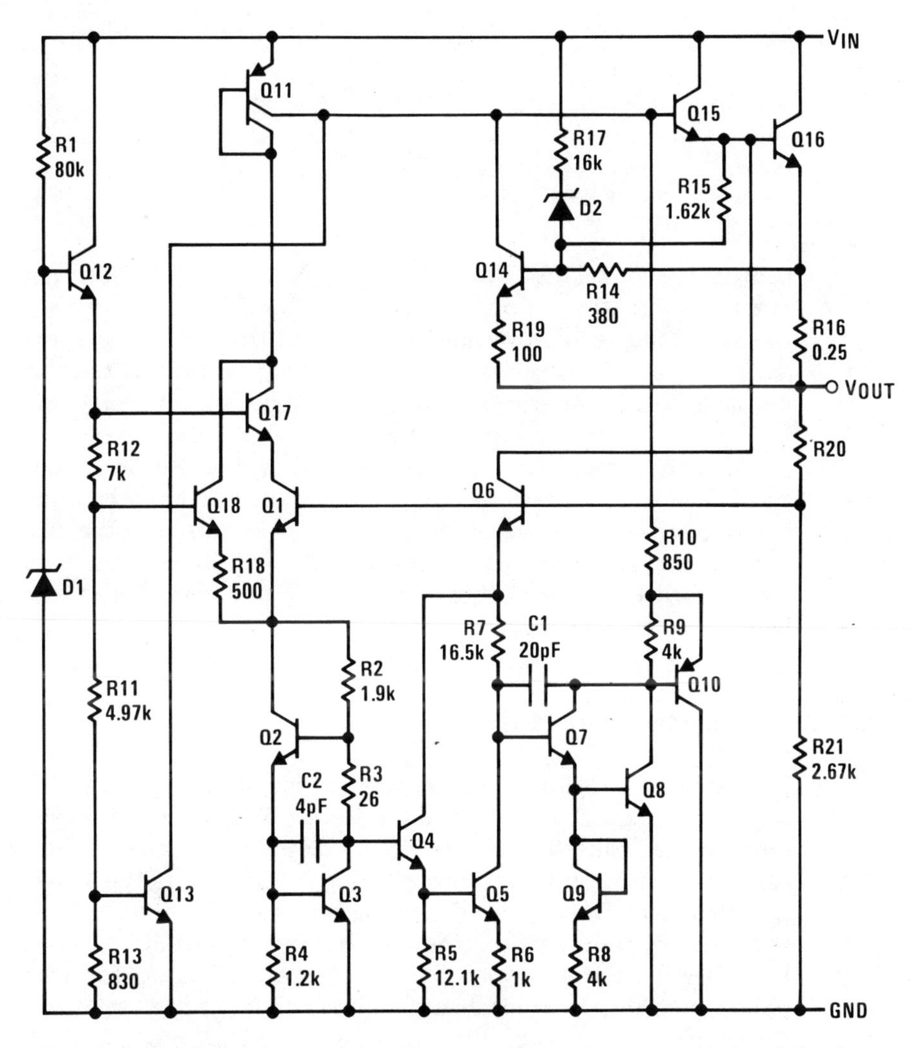

Figure 9-17 Technical data for a packaged regulator. (Courtesy National Semiconductor Corporation.)

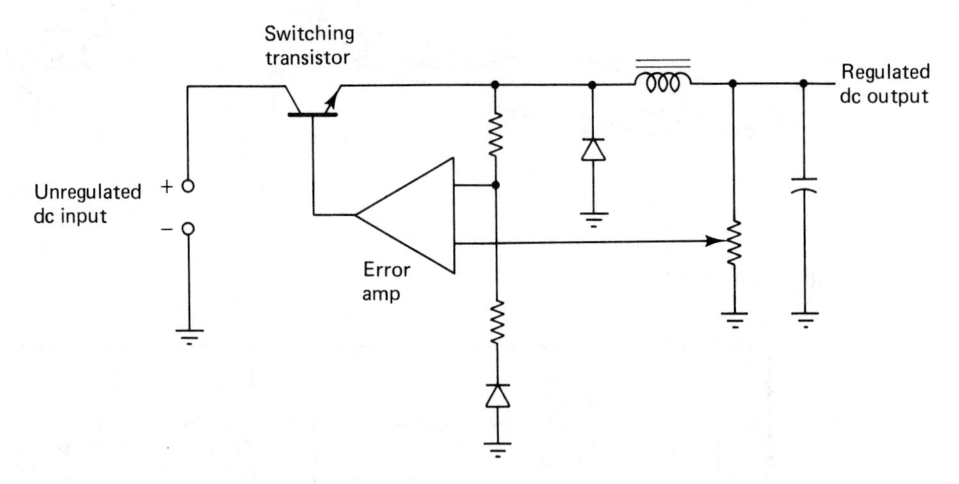

Figure 9-18 Schematic diagram for a switching type of voltage regulator system.

control the switching transistor operating cycle. A pulse is developed at the output due to the action of the two diodes. This square wave is used to turn the switching regulator transistor on and off. The output of the transistor is filtered and thus is available as a pure dc voltage at the output terminals of the circuit. Variations in output voltage are corrected in the error amplifier in order to maintain the desired level of voltage.

The four regulator circuits described in this section are all used in stereo units. There are variations for each basic circuit. In some instances the number of variations could exceed the pages in this chapter. The intent is to explain the operation of a basic system. The technician then has to apply this knowledge to the other similar types of circuits in an effort to comprehend how they function.

SERVICING THE POWER SUPPLY

Diagnosis and repair of the power supply section of a unit can be a very simple procedure when certain rules are followed. Before any test equipment is used, one should attempt to classify the type of problem. This can usually be done by considering these questions. Is the power supply completely dead? Do any of the indicating lights light? Is there some output from the system, even if it is not the proper output? Is any 60- or 120-Hz hum heard at the output of the system? Looking for answers to these questions will help to localize the power supply problem.

Refer to the block diagram shown in Figure 9-19. Consider the answers to the questions just proposed and how they apply to the block diagram of the power supply. If the indicator lights are lit, the input to the transformer

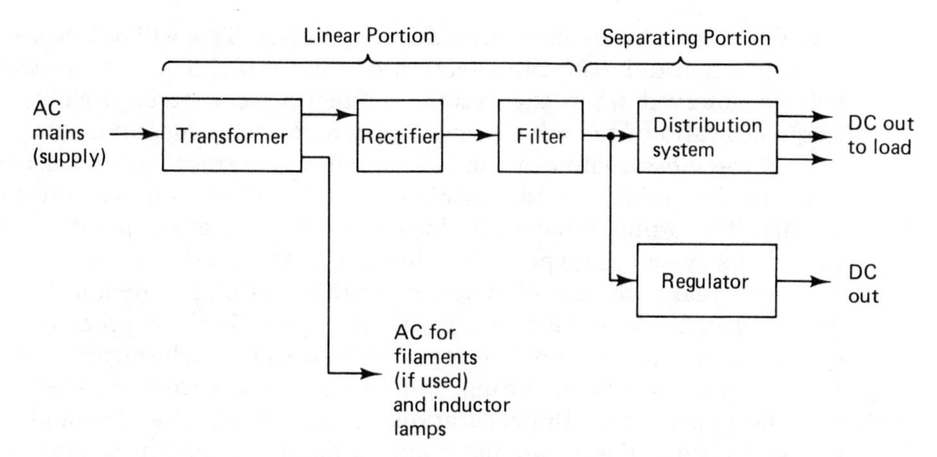

Figure 9-19 The blocks in a power supply can be divided into a linear section and a separating section for troubleshooting purposes.

is also working. This will usually eliminate the power cord, fuses, and transformer primary as probable areas of trouble. One does not have to use any test equipment to make this test.

A condition of low output and or low-frequency hum is usually related to the rectification, filtering, or distribution sections. At this point either a voltmeter or an oscilloscope should be used. The system is considered to be a linear flow path system. The rules related to this type of system are now applied. First, check the input to the system to make certain that it is correct. Use the information found on the schematic diagram for a reference. Next, check the output. This will be at the input point of the distribution system. One should expect to find the proper level of operating voltages and less than 1.5% ripple voltage.

It is necessary to know what to expect to find when making a measurement. For instance, the correct polarity and frequency of the voltage is critical. The frequency of the ripple voltage is dependent on the type of rectifier system used. Half-wave systems have the same frequency as the applied voltage. Full-wave systems will double the frequency of the output voltage. Typical waveform for both systems are illustrated in Figure 9-20.

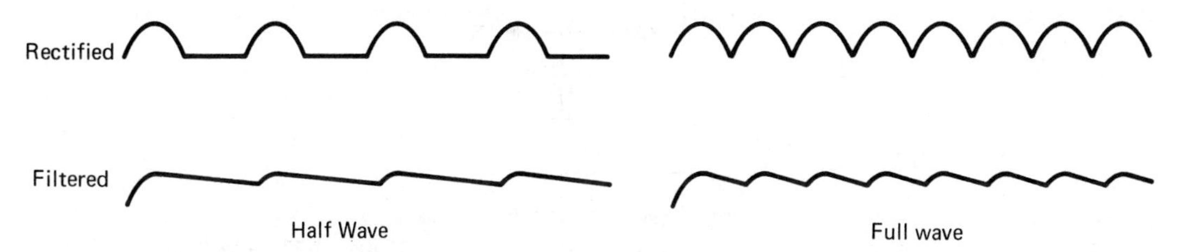

Figure 9-20 Waveforms found in half-wave and full-wave power supplies. One should expect to observe these when measuring with an oscilloscope.

The upper waveforms show an unfiltered output. This will not be seen unless there is a failure in the filter section of the supply. The bottom waveforms will be observed when the system is functioning correctly. Note, also, the frequency ratio of the waveforms. This is a necessary measurement.

If the measurement made at this point is correct, all systems from the input to this point are also working properly. There is no need to test them further. The problem has to be located between the test point and the output of the system. The procedure for testing this section is done by applying another signal path rule. A diagram for this portion is shown in Figure 9-21. This is called a *branching* or *separating system.* The test procedure for this system is first to measure the input. If it is valid, each output is measured. If all of the outputs are wrong, something is not correct between the input and the point where the separation occurs. In this case it would be either C_4 or R_4 since these two parts are common to the entire system. If one output is wrong, the problem is limited to being between the point of separation and the specific output that is wrong. If, for example, output C is wrong and all of the others are correct, the problem component is either C_2 or R_2, or possibly both. The test procedure used is to move the tester closer to a valid operating point with each test until the area of trouble is limited to a very small section of the unit.

Problems in the rectification and filtering section are usually related to the failure of rectifier diodes and filter capacitors. Both can be identified

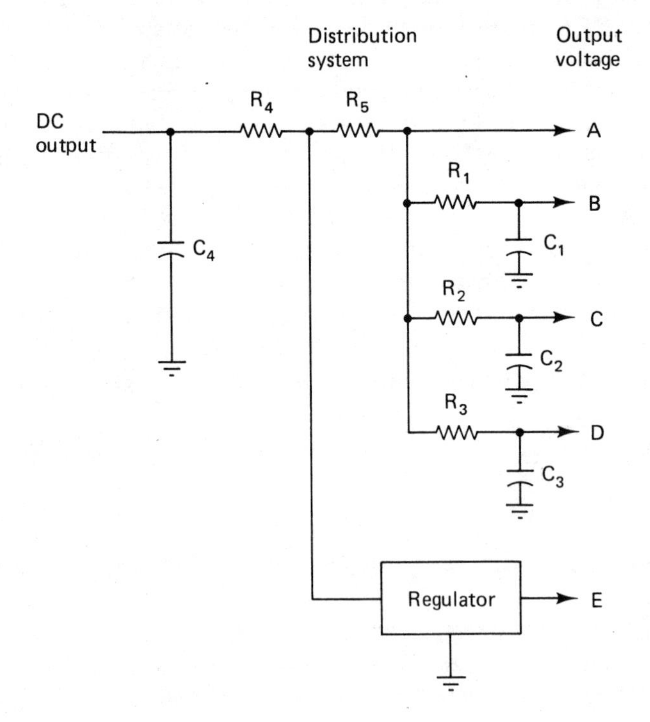

Figure 9-21 A voltage distribution system uses separating flow paths.

with the oscilloscope. Failure of the single diode with half-wave supply will either provide an ac voltage at the output or there will be no voltage at all. Failure of one diode in a full-wave system develops a half-wave output waveform. Failure of the filter capacitor will be observed as a nonfiltered output if the capacitor is open. If the capacitor shorts, the output voltage drops to almost zero and usually some other part will fail.

A word of caution is necessary at this time. Be sure that all the measurements being made are referenced to the proper point in the circuit. Often, the failure to use the proper reference or common point will provide invalid readings. An ac wave will be displayed on the scope when one fails to connect the common lead to the system. Be sure that both test leads are used and used properly at all times.

Another type of failure related to power supplies is the failure of the regulator circuit. A zener diode will open if too much current flows through it. The output is then unregulated. This may be caused by an overcurrent condition in the load. This is tested by measuring the dc resistance of the load and determining its current and power requirements.

Any regulator circuit will fail to operate correctly when the load demand exceeds the design of the regulator. Component and system analysis must consider this fact. The problem area has to isolated. In some instances the connections between load and power supply have to be separated to determine which system has failed. Often, a resistance measurement (with power off, of course) will help to isolate the problem area. Use of the schematic diagram will help to evaluate where the overcurrent point is located.

One last point about troubleshooting that applies to any circuit or section of all electronic devices is that resistors do fail. Usually, the failure is due to an overcurrent problem in the unit. The failure life of a low-power resistor is rather unique. When first overheated by excess current, the resistance value will drop to below its rated value. As the overcurrent condition continues due to component failure or the new low value of the resistor, the resistance goes lower until the elements separate, producing an open circuit. Under some operating conditions the resistance value will rise and develop a higher than normal value instead of a complete failure. This reduces current flow in the circuit and changes the operating conditions of the device.

Approaching the analysis of the system in an orderly fashion will aid in the rapid diagnosis and repair. Each part of the system can be treated by use of the signal-flow-path approach. This is also true for current flow paths; they can be analyzed in the same manner. The use of this approach may seem to be time consuming and slow at first. As one uses it and develops the habits required for proper analysis, the time for this is reduced and the efficiency increased. It is wise to develop the habit of "feedback" after completing a repair. Ask yourself: "How could I have done a better, more efficient job?" Use the answer to improve your techniques.

QUESTIONS

9-1. Draw a block diagram for a power supply that has both a regulated and an unregulated output.

9-2. What is the purpose of a power supply transformer?

9-3. What mathematical factor is used to convert ac rms values to the approximate dc value in a power supply?

9-4. What is the purpose of a filter capacitor?

9-5. What is the frequency of each of these power supplies: half-wave, full-wave center tapped, full-wave bridge, and half-wave voltage doubler?

9-6. Why is it important to observe the polarity of an electrolytic capacitor?

9-7. Why is a switching power supply used?

9-8. What is a zener diode, and how does it work?

9-9. Where is a good point at which to make the first test in a power supply?

9-10. What is the advantage of using an IC regulator instead of discrete components?

Chapter 10

Audio Power Amplifiers

The main purpose of any audio output circuit is to develop sufficient power to operate the speaker system. Speakers operate on electromagnetic principles. A low-voltage, high-current power source is required in order to have the cone move. This power source is developed in the output stage of the audio amplifier. There are several methods of developing the proper amounts of voltage and current. These include use of either an audio output transformer or a coupling capacitor in order to transfer this energy.

The energy source for the output stage is developed in the stereo amplifier's power supply. This supply is designed to provide the proper levels of voltage and current. These values are used to provide operating power to the output stages. The signal that is fed to the output stage is used to vary the voltage and current flow in the system. It, in a sense, is the control section of the system. An amplifier is a device that uses a small amount of power to control a larger amount of power. The signal acts in this manner.

This chapter explains the operation of several types of audio output systems used in stereo service. In addition, some of the problems commonly found in audio output systems are described. Methods of troubleshooting these systems are also presented. Some of the basic theories about power amplifiers were presented in Chapter 4. The material in this chapter is more detailed.

GENERAL TROUBLESHOOTING PROCEDURES

Experience and research in the servicing of electronic products led to the conclusion that those components that work the hardest are often the ones that fail first. This is very true when one services the audio amplifier. The hardest-working components in the amplifier are the output transistors. Their failure rate is much higher than that for any other components in this unit. When localizing a problem in the audio system one should follow the general rules for troubleshooting. A visual inspection of the circuits and circuit boards is done *before* any test equipment is used. If one sees burned or broken components, the servicing should be focused on that area first. It is often not necessary to follow the system established for a linear signal flow path when obviously defective components are found in the circuit.

However, it is easy to jump to conclusions when servicing a unit with burned parts. The inexperienced technician will immediately replace the bad components and then test the repair. The major problem with this approach is that the *cause* of the problem often is in another area of the unit. This is very true in direct-coupled amplifiers. A partial schematic for a dc amplifier is shown in Figure 10-1. It is the vertical amplifier of an oscilloscope. All stages are directly coupled to the push-pull output circuit. When the input transistor, Q_1, is shorted, every operating voltage in the amplifier is thrown off. One output transistor becomes very hot and often fails. Changing it will replace the failed component but will not correct the problem. A circuit of this type requires further diagnosis. The best way to localize a problem of this type is to measure operating voltages and compare them to those provided by the manufacturers. Often, a defect in the input circuit of a directly coupled amplifier will cause all voltages to be wrong. Looking at the beginning of the system is a logical starting point in this situation. This type of situation is usually true only for directly coupled stages.

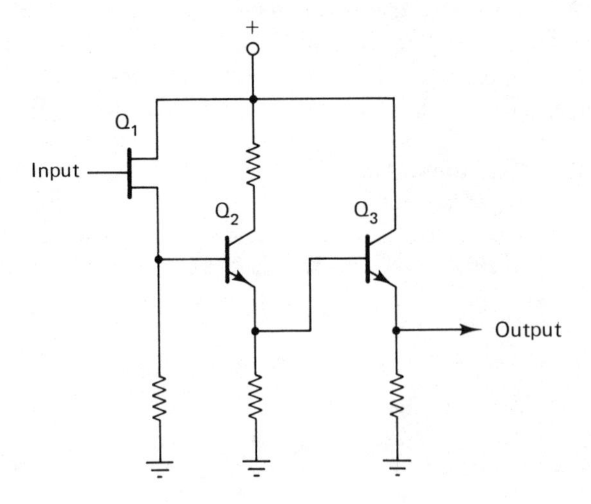

Figure 10-1 Partial schematic for a direct-coupled amplifier system.

When one cannot see a physical problem and uses this to localize the trouble area, the general rules for troubleshooting are applied. The system is approached as a linear signal path system. A valid input signal is developed. This is fed to the audio input of the amplifier. The problem area is localized using the half-split method. Continually make a test at or near the middle of the system. Each test should eliminate one-half of the remaining blocks. This will reduce the area of trouble to one stage. The material in this chapter assumes that this method has localized the area of trouble to the output stage of the amplifier.

Use of a load. The effective technician will use knowledge and equipment to repair a problem in the unit. This requires thought and planning before the actual tests are performed. One part of the setup procedure that is often overlooked is the use of a load connected to the output of each channel. This load may be in the form of a fixed resistance. Several manufacturers make resistors with low resistance and high wattage ratings. These types are shown in Figure 10-2. These resistors should match the output impedance of the amplifier. This value is 4, 8, or 16 Ω. In addition, the wattage rating for the resistor should be higher than that of the amplifier being tested. One resistor is required for each channel of the system. Two

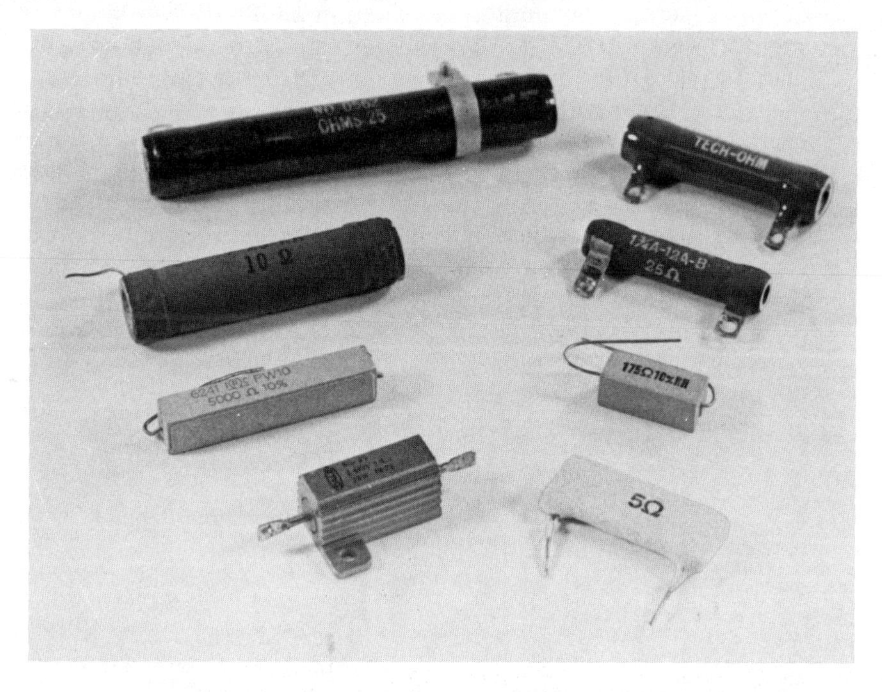

Figure 10-2 Various power resistors can be used for loads when testing power amplifiers. (From Joel Goldberg, *Fundamentals of Electricity*, © 1981, page 31. Reprinted by permission of Prentice-Hall, Inc.)

8-Ω resistors could be used for each channel. When these are series wired, the effective resistance is 16 Ω. When they are parallel wired, the effective resistance is 4 Ω. A load resistance with a wattage rating that is above the requirements of the amplifier may be used safely with low-power amplifiers.

Second-channel tests. The technician is very fortunate when servicing a stereo amplifier. There are two identical channels in the system. The voltage and resistance values in the working channel can be used to help locate problems in the nonworking channel. In addition, waveforms from one channel can be compared with waveforms found in the other channel. A dual-trace scope is used to compare both the phase and amplitude of the audio signals at similar points in each channel.

Voltage tests. Once the area of trouble is reduced to the audio output stage, the next approach is to measure the voltages in the system. Start with the voltage from the power source. If this is low, it could influence the operating conditions of the amplifier. After determining that these values are correct, the next voltage measurements should be at the elements of the output transistors. Compare these with those on the schematic or taken from the working channel. One good test is the measurement of the voltage that develops across the emitter-to-base junction of a working transistor. This should be about 0.7 V for a silicon transistor and about 0.2 V for a germanium type of transistor. The polarity of the transistor has no effect on this value. Both NPN and PNP transistors, as shown in Figure 10-3, have this characteristic. It should be close to the voltages given. If it is higher than the correct value, the transistor is leaky and must be replaced. A lower voltage reading indicates a shorted transistor. It, too, must be replaced.

Another test, which is valid only for series-connected output transistors, is to measure the voltage at the midpoint of the two transistors. This should be done with no signal applied. This voltage should be about one-half of the applied voltage in the system. Do not overlook the bias voltage developed for the transistors. This is usually developed from a bias network. Its failure will cause a malfunction in the system.

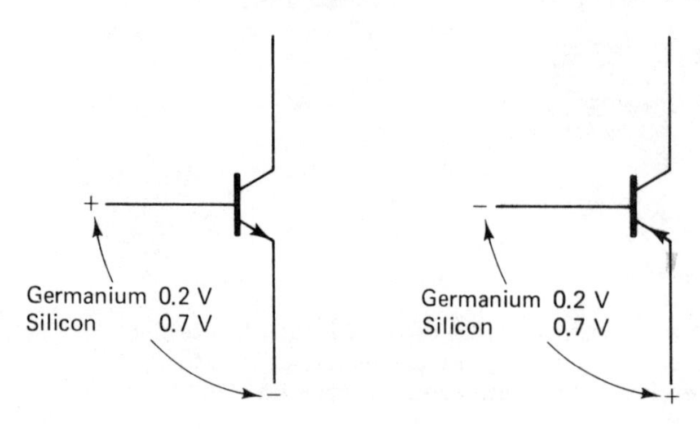

Figure 10-3 Operating voltage polarities for NPN and PNP transistors.

Resistance tests. Another way of testing an output stage is by making resistance measurements. These are always made with the power off in the system. It may be necessary to discharge electrolytic capacitors in the power source or to isolate components in the circuit in order to obtain a valid measurement.

It may be necessary to isolate the specific circuit from the rest of the system when making a resistance test. This is done as shown in Figure 10-4. The only practical method of separating an etched circuit is to cut the conductive path. Use a sharp razor or knife to separate the foil pattern. A small cut is all that is necessary. Cut the foil as shown between the area in which resistance is being measured and the rest of the system. After all the tests are made, this separation is repaired by soldering a small piece of wire across the open spot in the foil path.

Be sure to check the ohmic value of bias resistors. Often, an overcurrent situation will raise the resistance of the unit without discoloring or breaking the resistor case. When a bias resistor increases, look for a transistor that is also defective. It is not a good idea to "shotgun" a circuit by replacing all

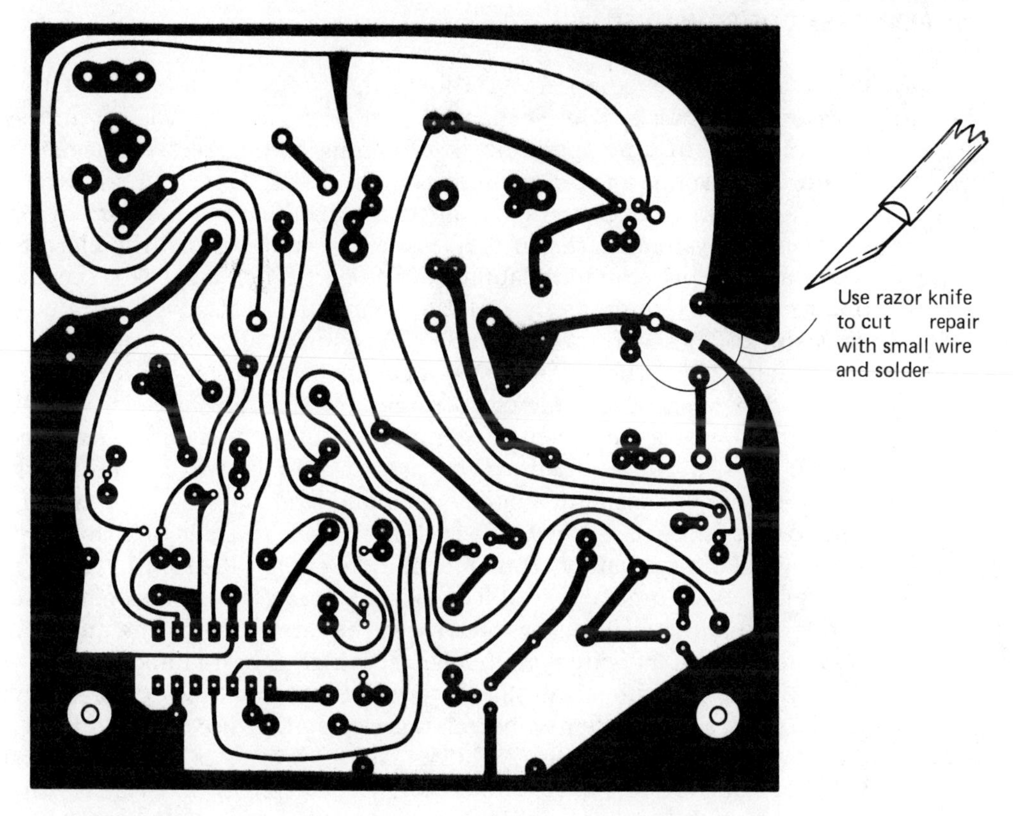
Use razor knife to cut repair with small wire and solder

Figure 10-4 Often a circuit foil path has to be cut in order to isolate the circuit from the rest of the system.

components. There are times, however, when the experienced technician replaces specific "extra" parts. This is true when output transistors are replaced. It is a good idea to replace the emitter-bias resistor. These resistors have a reputation for failing just after a new transistor is installed. In this situation it is better to replace this resistor and the output transistor as a precautionary repair. Recalls are both expensive and time consuming. In this situation a little extra precaution pays off.

Oscilloscope testing. This is the only common electronic test instrument that will display a voltage waveform. Once it has been determined that the input wave is valid, the output wave is tested. One needs to know the type of amplifier system used and what typical waveforms should look like. If the waveforms are not faithful reproductions of the audio wave, the area of trouble is located. The amplitude values are also important. This depends on the type of amplifier circuit. This factor is discussed later in the chapter.

REPLACEMENT OF COMPONENTS

When a defective component is located, it must be replaced. This is not always an easy task. The mechanical part of replacement is often easy. The difficult part of the operation is obtaining the correct replacement part. There are several sources of replacement parts. One source is the manufacturer of the system. Almost all companies have some sort of national distribution system. Often, it is necessary to write or call the closest outlet to obtain the price and availability of the part. Then a check is sent with the order. Another source is one of the companies that advertise in electronic magazines. These companies often have original parts available. This is particularly true when one needs Japanese transistors. A third source is a local independent electronic parts distributor. These companies usually carry a line of universal replacement components. Often these are exact replacements, but in some circuits the parts recommended do not work well. Experienced technicians recognize this and will be very selective of which components they use. The whole system becomes a trade-off. Universal parts are available immediately from a local independent distributor. They usually cost more than orginal parts. The original parts are not as convenient to obtain and take much longer to arrive at the service shop. The question to be answered by the individual technician is: Which is more important?

Another consideration that is very important is the correct part number on the device. Japanese numbered transistors often use a letter following the part number: for example, 2SC321*A*. The letter *A* is used to indicate the beta, or gain factor, of the transistor. Other transistors will use the same number system but have a different letter. When replacing transistors of this type, always select one that has the same letter as the one being replaced.

Failure to do this will establish a circuit that has incorrect gain and may throw off the proper amplification in the system.

Heat sinking. Another factor that is very important is the use of a proper heat sink. The design of the system often includes a heat sink for the power transistors. The purpose of the heat sink is to remove heat from the transistor package. This permits the transistor to operate at higher than normal temperatures and is directly related to the power output of the device. The heat sink is actually a heat transfer and radiating device. The surface area of the sink is designed so as to transfer as much heat as possible into the atmosphere. Do not attempt to operate any solid-state device without its heat sink for an extended period of time.

A second factor to consider when replacing a semiconductor that is sinked to a metal surface is the insulating washers and fasteners, shown in Figure 10-5. Many semiconductor devices have one element connected to a heat-sinking surface. This is usually the collector element of a transistor. In many circuits the collector is not operated at zero potential. The use of an insulated washer between the semiconductor and the heat sink will permit heat transfer without any electrical connection. Some installations also use a plastic nut and bolt for the fastener in the system.

A third factor to consider when replacing a semiconductor in a heat sink is the use of a silicon grease. This grease is a nonconductor of electricity. It is an excellent conductor of heat. Proper use of silicon grease on both surfaces of the semiconductor and heat sink is required for maximum heat transfer. A word of caution is necessary about silicon greases. There are some transistors that are enclosed in a flat plastic case. These cases are subject to failure if one uses white silicon grease. The breakdown does not occur immediately, but does occur over a period of time. Use clear grease for all applications and this will not be a problem. Now that the generalizations about power amplifiers and specific information about replacement are explained, let us look at specific circuits.

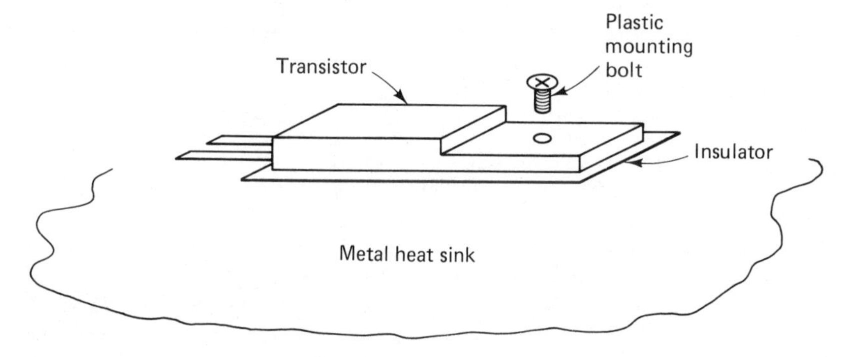

Figure 10-5 Power transistors require an insulated spacer when their collector is not at circuit common.

TRANSFORMER-COUPLED SYSTEM

One method of developing power to operate speakers is the use of an audio output transformer. This transformer is used as an impedance-matching device as well as for the transfer of power. A typical audio output circuit for one channel of a stereo amplifier is shown in Figure 10-6. In this circuit transformer T_1 is used as an input transformer. The transformer has a single primary and a center-tapped secondary. The secondary of this transformer acts as a phase splitter in this circuit. When the input signal is positive at the upper lead of the primary, the signal voltage developed at the upper winding of the secondary is negative. This makes the lower end of the winding negative with respect to the center-tapped common connection.

Both transistors used in the output stage are NPN types. Both will conduct only when the signal goes positive. Resistor R_1 is used to bias these two transistors at, or near, the point of cutoff. They operate as class AB amplifiers in this circuit. When signal to Q_2 is positive, it will conduct. This develops a current flow through it and one-half of the primary of the output transformer T_2. The current flow through this half of the primary develops a voltage across the secondary winding. The resulting waveform is one-half of the sine wave used as the input signal. This is illustrated in Figure 10-7. The

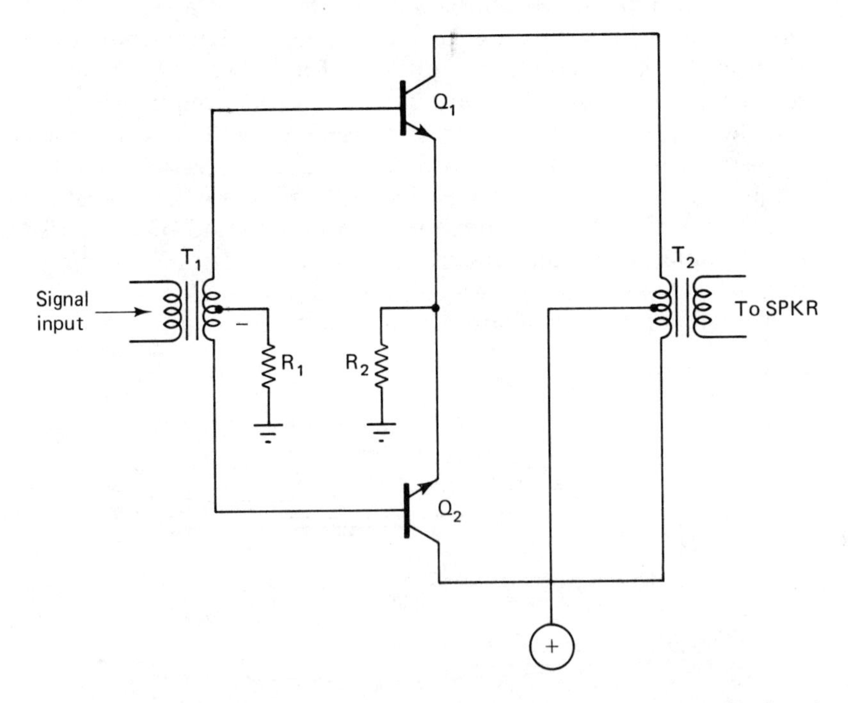

Figure 10-6 Schematic diagram for an audio output stage using input and output transformers.

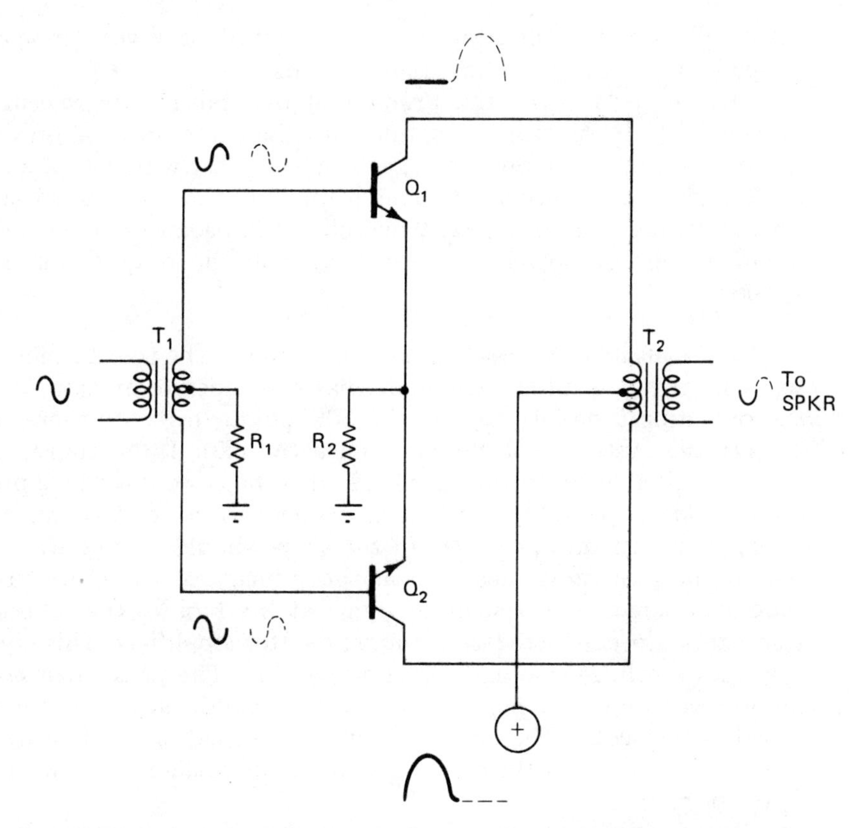

Figure 10-7 Waveforms found in the transformer-type audio output system.

solid lines on the sine-wave forms represent the conduction that occurs during the first half of the input signal cycle.

When the input signal goes negative, all the signal polarities in the system are reversed. These are shown by dashed lines in the drawing. This reversal turns off transistor Q_2 and turns on transistor Q_1. It now conducts and a current flows through the other half of the output transformer primary. This current also induces a voltage in the secondary. This voltage has a polarity that is opposite to that created during the first half of the cycle. The result of this action is the recreation of the sine wave at the output terminals of the transformer.

What is happening in this circuit is that each of the output transistors is being turned on during one-half of the cycle of the input wave. A current flows through half of the primary of the output transformer during each half of the cycle. The output winding of this transformer is not center tapped. This means that all the current in the secondary flows through the speaker's voice coil. The electromagnetic field produced in the speaker causes the voice coil to move. The result is a movement of air waves. These are then received by the ear and translated into sounds.

The impedance of the primary of the transformer is usually on the order of 100 Ω or more. This matches the impedance of one of the two transistors used in the output stage. The secondary winding for an audio output transformer is designed to match the impedance of the speaker system. This is usually 4, 8, or 16 Ω. When all the impedances are correct, a maximum amount of power is transferred from the output transistor to the speakers.

Troubleshooting the transformer output system. The transformer output system uses two transistors. These are biased to operate in class AB. This puts the operating time just beyond the 50% point. In reality there are two circuits in this system, with a common negative path. If one transistor fails, the other is still able to funtion as it did when both were working properly. An oscilloscope is probably the first instrument to use to test this circuit. The waveforms seen on the screen of the scope should be sine waves. The output of the transistors measured at the primary of the output transformer should be larger than the input signal at the base of the transistor. These transistors are connected as common-emitter amplifiers. This configuration produces a voltage gain and a phase inversion. The phase inversion may not be observed on the scope. This depends on how the sweep triggering is established in the scope. The audio output transformer is a step-down type. The sine wave observed at the secondary should be smaller than the one observed at the primary.

If one fails to observe the necessary voltage amplification in the transistor amplifiers this is probably the trouble area. A voltmeter can be used (or a scope measuring dc volts) to measure the voltage developed between base and emitter. If this voltage is not about 0.7 V the transistor is leaky and needs to be replaced. The only other components that could be a problem in this circuit are the bias resistors and the transformers. Resistance measurements of both will determine if they are defective.

COMPLEMENTARY-SYMMETRY SYSTEM

A second method of developing the required amount of output power uses a complementary pair of output transistors. The term *complementary pair* refers to two transistors, one NPN and one PNP, that have identical operating characteristics except for their polarities. These two transistors are used in output circuits. The advantage of their use is that the output transformer is no longer required. A schematic diagram for this type of circuit is shown in Figure 10-8. The two transistors are connected in series. Both are emitter-follower or common-collector amplifiers. This means that the voltage gain in

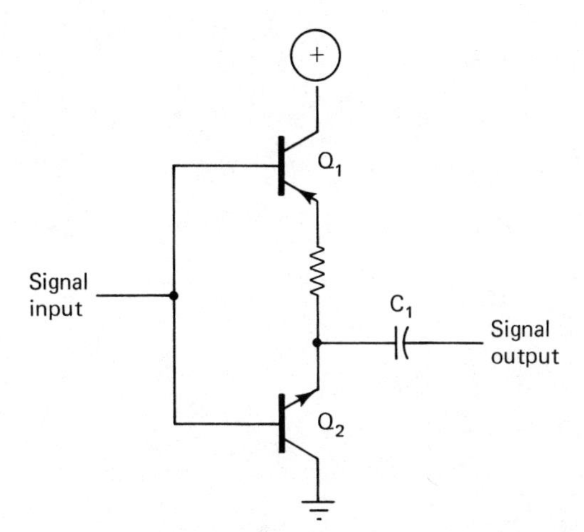

Figure 10-8 Schematic diagram for an audio output system that does not require transformers.

the amplifier is less than 1. It is about 0.98 of the input signal. The advantage of this system is that each transistor operates for slightly more than 50% of the duty cycle. The signal input to the two transistors is a common connection from the preceding stage. Since the two transistors are of opposite polarity, they operate during only one-half of the input wave cycle. Each transistor acts as a controlled resistance in this circuit. Since their emmiters and collectors are wired in series, they act as two variable resistances. This is illustrated in Figure 10-9. The output of the circuit is the midpoint connection between the two transistors. This is a dc voltage point. The specific value of this voltage depends on the dc resistance of each transistor. When they have no signal applied, their resistance is equal. The voltage at this time is equal to one-half of the source voltage. When transistor Q_1 is turned on by a signal, its resistance decreases. This changes the voltage drop relationship of the series-wired transistors. The dc output voltage will rise as Q_1 conducts.

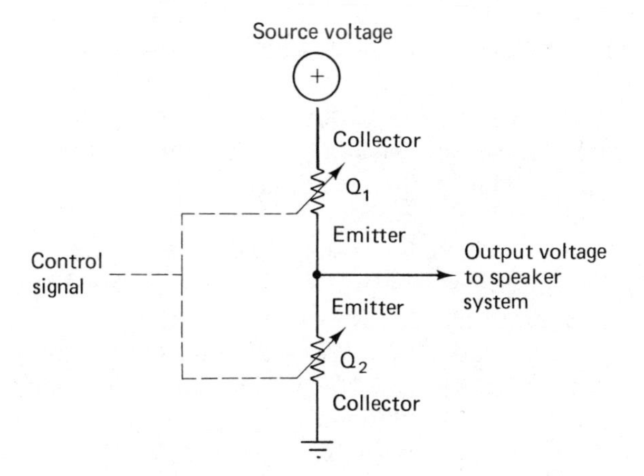

Figure 10-9 Operation of the system using variable resistors instead of transistors as an example of how the system works.

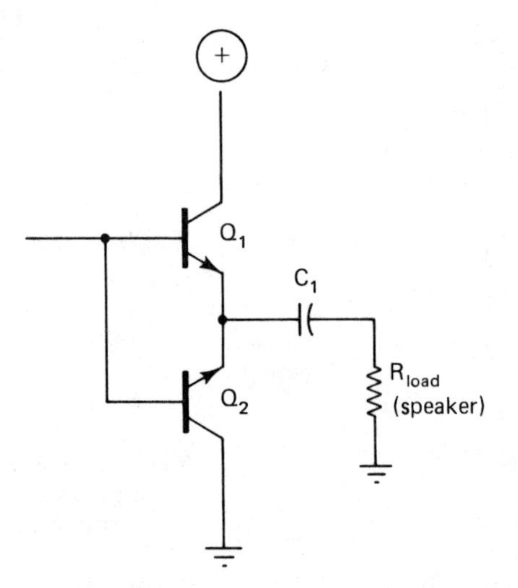

Figure 10-10 Schematic diagram for a complementary-symmetry audio output system using NPN and PNP transistors.

Its resistance changes are directly related to the amplitude and shape of the input signal.

During the second half of the input wave cycle, transistor Q_2 is turned on. This decreases its interval resistance and reduces the voltage measured between the output point and circuit common. What now occurs is that the dc voltage at the midpoint, or output, connection is varied as the input signal causes each transistor to turn on. This is an alternation system because each transistor operates during only one-half of the input signal cycle.

The purpose of this system, or any other output system, is to develop a current flow through the load. The output circuit shown in Figure 10-10 shows how this is accomplished. The varying dc voltage on the left-hand plate of capacitor C_1 will change the charge on this plate. This, in turn, will change the charge on the right-hand plate of the capacitor. The speaker voice coil is connected between common and this capacitor plate. The charges in the capacitor produce a source for electron flow. This causes a current flow through the load and develops its magnetic field. The magnetic field reacts with the permanent magnet in the speaker. The end result is the movement of the voice coil and speaker cone in order to produce sound waves. There is a fairly high current requirement for this action. To accomplish this, the coupling capacitor, C_1, must be rated at 1000 μF or better.

Troubleshooting the complementary-symmetry system. The most practical method of troubleshooting this circuit starts by first measuring the dc voltage that develops at the output (emitter) connections of the transistors. This voltage should be about one-half of the total voltage applied to the circuit. Should this voltage be different by more than 10 to 15%, one of the transistors is probably defective. A voltage measurement between base and emitter will determine if this is true. An oscilloscope can also be used for these tests.

An analysis of the input and output waveforms of the circuit will show if it is working properly.

It is not always feasible to operate the circuit at full power. There are times, particularily after replacing the output transistors, when the technician needs a better control of the circuit. When this is true, a variable autotransformer is used to bring the operating voltages up to correct levels. The emitter-to-base junction of one transistor can be monitored. Start the transformer at its lowest output and connect it to the ac power input of the system. Slowly advance the transformer control until the transistors are turned on. This point is where the base-to-emitter voltage reaches the operational level of 0.7 or 0.2 V. If all systems are still working properly, the transformer output is increased until it reaches normal operating levels. If the operation of the system starts to shift into an abnormal condition, the transformer output can be quickly reduced. This will save the power transistors from destruction.

The only other problems that occur in this system relate to any emitter-bias resistors and to the coupling capacitor. Measure the resistance of the emitter bias resistor. An overcurrent condition will often cause this resistor to increase in value. Should the coupling capacitor fail by opening, there is no output to the speaker. If this capacitor shorts, the dc voltage at the emitter junction is connected directly to the speaker. This often causes an overcurrent in the speaker system and a resultant failure in the voice coil of the speaker.

The expereinced electronic service technician will replace the output transistors *as a set*. Very often a defective output transistor changes the operating conditions of its opposite unit in the system. The changes are not always noticeable. When the obviously defective transistor is replaced and the unit turned on, the other transistor can either fail or make the replacement transistor fail. It is wise to order and install a matched pair of transistors in order to minimize this problem.

BRIDGE OUTPUT SYSTEM

The final system to be described is called the bridge output system. It is very different from other systems. The schematic diagram shown in Figure 10-11 shows this system. It is actually two complementary-symmetry output systems with the speaker connected between the midpoints of each system. The two systems are fed signals that are out of phase with each other. This produces a rise in the emitter voltage on one side while the other side has a drop in voltage. The differences produce a current flow through the speaker voice coil.

The bridge output circuit is often manufactured as a solid-state module. There are no discrete components, only one module. Typical modules used

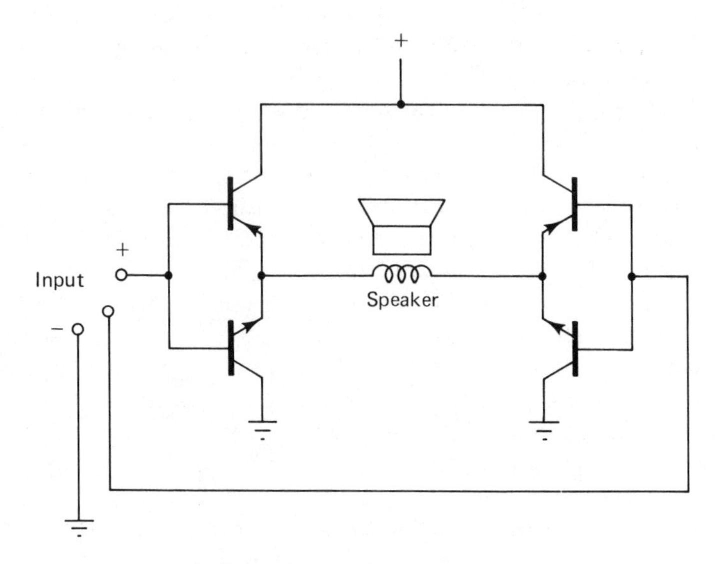

Figure 10-11 Schematic diagram for a bridge audio output system. Both speaker leads are above circuit common in this system.

for audio output systems are shown in Figure 10-12. These modules are available in several sizes and forms. As a rule, the larger modules are able to produce a higher power output for the system. These modules are used in large stereo systems and in smaller units, including automobile radios. Usually, one channel's output system is contained in one module. Two modules are required for a stereo system.

Troubleshooting the bridge output system. The easiest way to test this system is with a voltmeter or an oscilloscope. Measure the dc voltage at the emitter connections. This should be one-half of the supply voltage. Any other value indicates an imbalance and usually a defective component. When

Figure 10-12 Audio output modules containing a complete output system for stereo amplifiers are made in different sizes.

testing a module, the first test is to check input and output signals. If there is no output signal, measure the operating voltages as given on the service literature. Wrong value voltages are indicative of a bad module, as is no output voltage at all. The only solution to this is to replace the module with a known good one.

There are other instances where there is a short circuit inside the module. This can be verified by a resistance check of the output connections. It may be necessary to cut the foil on the circuit board to verify the short circuit. The only solution is to replace the module.

Servicing audio output stages is not very difficult. The major problem is that too many technicians are parts changers instead of troubleshooters. Learn to look for reasons as to why a circuit does not work properly. Use the knowledge of how a series circuit is supposed to function for your analysis. After all, the output circuit of a push-pull amplifier is not much more than a series circuit! Learn to reduce the large systems to smaller subsystems. Then it is much easier to apply those rules developed by Kirchhoff and Ohm. This habit will help you to develop into an experienced, efficient technician.

QUESTIONS

10-1. What is the source of energy used in a power amplifier?

10-2. Where is a suggested place to make the first test in a dc amplifier?

10-3. Why is it necessary to use a load when testing a power amplifier?

10-4. What are second-channel tests, and how are they used?

10-5. What advantage is there to using general replacement transistors instead of exact number replacement?

10-6. What are the disadvantages of using general replacement transistors?

10-7. What is a heat sink, and why is it used?

10-8. Briefly describe the troubleshooting procedure for a complementary-symmetry system.

10-9. Why should output transistors be replaced as a matched set?

10-10. What cautions should be used when working with bridge output circuits?

Chapter 11

Low-Level Amplifiers

The term "low level" is used to describe low-power amplifiers. These are the stages of an amplifier that are used to amplify the signal from the input transducer. There are usually several stages of amplification. These stages develop the signal to the level where it is used to drive the power output stages. Amplifiers are classified in several different ways. One method is to relate the amplifier to its position on the block diagram. For example, input amplifier stages are often called *preamplifiers*. These stages take the input signal from the transducer and provide voltage amplification. The stages that follow the preamplifiers are called *voltage amplifiers*. These stages provide additional signal voltage amplification. No matter what the name of the stage, its basic purpose is to provide signal amplification.

There are many methods of accomplishing the necessary amount of amplification. The most common method is the use of a voltage amplifier transistor stage. The circuit usually used for this is called a common-emitter amplifier circuit. This circuit has more desirable impedance matching and power transfer characteristics than other types of amplifier circuits. Its characteristics and those of other amplifiers are discussed in this chapter.

Amplifier classification by type can also be done using other significant factors. One such factor is the method of transferring signals from one stage to another stage. These classifications include direct, transformer, and capacitive coupling devices. Still another method of identifying amplifiers is by their usage. One can have a RF amplifier and an IF amplifier in addition to an audio-amplifier stage. Even though all the terms identified are correct, it is very unusual to use all of them to describe any one stage in a system. The stage may be a common-emitter transformer-coupled IF amplifier, but it is called an IF amplifier. The other terms are valid, but they are not often used in the description of the circuit. Coupling methods as they are applied to audio amplifiers are also discussed in this chapter.

BASIC AMPLIFIER CIRCUIT

Technicians are familiar with the term "amplifier." These same people usually are able to quote the definition for an amplifier. Few, however, really understand how an amplifier functions. A thorough understanding of the operation of a transistor amplifier will go a long way toward learning how to repair equipment correctly and rapidly. A look at Figure 11-1 will help in the understanding of how the transistor can amplify.

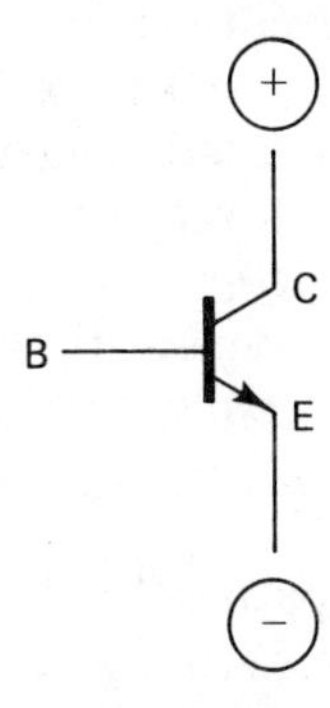

Figure 11-1 The transistor emitter-collector connections act as a variable resistor in a circuit.

Three elements are used in the junction transistor: the base, the emitter, and the collector. A NPN type of transistor is shown in the drawing. A PNP type will display similar characteristics. There are two conductive paths in the transistor. One of these is between the emitter and the base elements. The second path is from the emitter, past the base, and then to the collector element. The emitter and base elements are forward biased in an operational circuit. This means that a very low voltage will produce a high current flow across the junction of these two elements.

The emitter-to-collector junction, as it is called, is really from the emitter, through the base, and then to the collector. This path has a reverse-biased connection between the base and the collector elements. It takes a very high voltage to develop a current flow between these two elements. In an operational circuit the emitter element is the most negative element. The base is usually about 0.7 V above the voltage in the base. The collector has a much higher positive charge than any of the other elements. About 5% of the total current flow occurs in the emitter-to-base circuit. The balance, or 95%, occurs between emitter and collector.

What all of this means to the technician is that the emitter-to-collector current path acts as a controlled variable resistance in an operational circuit.

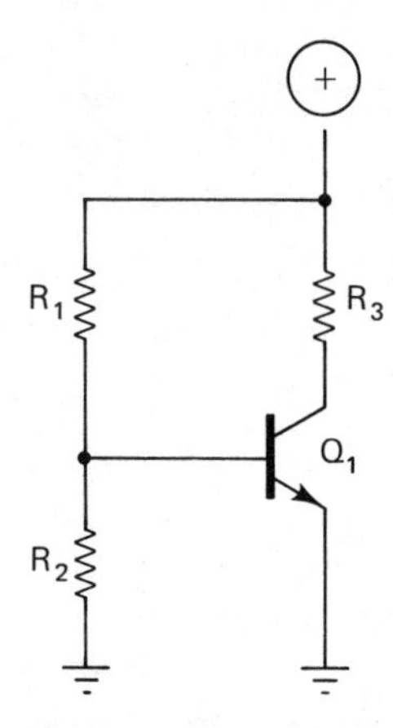

Figure 11-2 A transistor amplifier uses the emitter-to-base circuit to control the resistance between emitter and collector.

The emitter-to-base current is used to control the emitter-to-collector current. The circuit shown in Figure 11-2 will be used to help explain this. Resistors R_1 and R_2 are used in a voltage-divider circuit to establish the bias and operating point for the transistor. The voltage developed at their junction controls the emitter-to-base current in the transistor. This will be about 5% of total circuit current flow. The major current path in this circuit is that of the emitter to collector in the transistor and then through the series-connected load resistor R_3. The balance of circuit current flows through this path.

The major current path acts like two series-connected resistances. The resistance of the load resistor is a fixed value. The resistance across the transistor terminals is a variable value. Its resistance is determined by the bias voltage at the base of the transistor. The voltage at the junction of the load resistor R_3 and the collector of the transistor can change due to the amount of conduction that occurs in this circuit. This change in voltage can range from close to the value of the source voltage to close to zero. The exact value depends on the conditions of the control voltage in the base-emitter circuit. This voltage is used to control the conditions that exist between the emitter and the collector terminals. This action supports the statement that the input (base-to-emitter) circuit is used to control the output (emitter-to-collector) circuit. The action is also directly related to how a transistor, or a vacuum tube, acts as an amplifier.

The circuit shown in Figure 11-3 is very similar to the one in Figure 11-2. The difference is the addition of an input signal and the identification of an output connection. The sine-wave signal voltage is connected between the base and circuit common. Its voltage adds to that of the dc bias established by resistors R_1 and R_2. This additional voltage changes the operating point of the transistor. When the input voltage increases, the conduction of the transistor also increases. This acts as a reduction in the emitter-to-collector resistance. The result is a lower dc voltage measured between

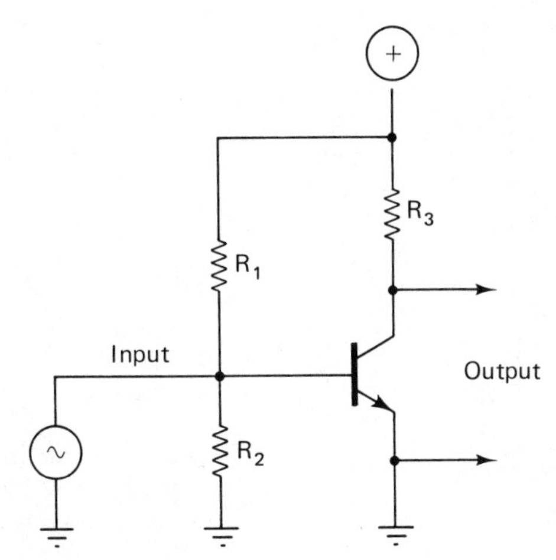

Figure 11-3 Transistor amplifier circuit described in the text.

the collector and circuit common. The opposite is also true. When the input voltage drops, the conduction through the transistor is reduced. This raises the dc voltage between the collector and circuit common. Another way of describing this action is to say that the input voltage is used to vary the dc resistance and the resulting voltage drop that develops across the emitter-to-collector terminals of the transistor. This principle is used in order to have transistors act as amplifiers. The three basic amplifier circuit configurations commonly used in electronic devices are described in the next section.

Common-emitter circuit. Transistor amplifiers are identified by the element used as signal common. This has absolutely nothing to do with the electron flow path. It should not be confused with the dc current path through the transistor. The signal path identification refers to the elements used for signal processing. One of the most common transistor amplifiers is called the common-emitter amplifier. This circuit is the same as that shown in Figure 11-4. The signal voltage is injected between the base and emitter elements. The output connections are made to the collector and emitter elements. In this circuit the emitter element is used for both input and output. This is why this circuit is called a common-emitter amplifier.

The operation of the common-emitter amplifier provides a fairly high voltage gain and a phase reversal of the signal. The input signal, shown in Figure 11-4, has a positive-going wave during the first half of its cycle. This will produce an increase in current flow in the transistor and reduce its collector-to-emitter resistance. The result of this action is a drop in the dc voltage measured between the collector and circuit common. The design of this circuit permits a much larger voltage drop at the collector than that used at the base to produce the drop. A 0.5 V change in base voltage could

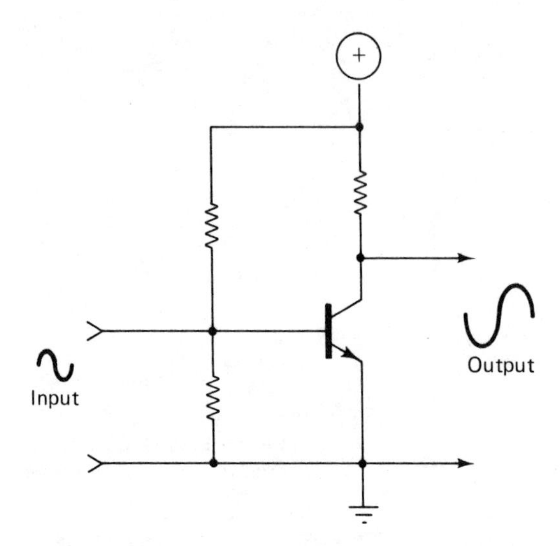

Figure 11-4 A common-emitter amplifier circuit has the input connected between the base and common and the output connected between the collector and common.

produce a drop in collector voltage and the resultant drop in collector voltage results in a signal phase reversal.

When the input signal goes negative, as it does during the current flow through the collector-emitter path, it causes an increase in internal resistance in the transistor. The collector voltage now rises. It rises to about the same value that it dropped to during the first half of the input wave cycle. This action is also out of phase with the input signal polarity. When an alternating signal is used at the input, the dc voltage at the collector changes at the same time. The changes are larger and they appear to have the same shape as the input signal except for their 180° phase reversal. This is the operation of the common-emitter amplifier. It uses a small signal injected between the base and emitter to control a larger dc voltage change between the collector and emitter. The changes in dc operating voltage at the collector look like the input signal except that they are bigger and are phase reversed.

One problem that exists with the common-emitter amplifier is called *thermal runaway*. This condition is caused by heat. Transistor operation will produce heat. The resulting heat will reduce the internal resistance of the transistor. This produces an additional amount of current flow and results in more heat. The condition continues until the transistor is destroyed. This problem is overcome by the use of an emitter resistor. The emitter resistor is usually a very low value, often less than 10 Ω. There is a negative effect to this circuit when the emitter resistor is used. This is called *degeneration*. Some signal develops across this resistor, as shown in Figure 11-5. The signal is opposite in phase to that developed at the collector. These two signals add and the result is a lower-than-desired output signal.

A method of resolving this problem is shown in Figure 11-6. A capacitor is connected in parallel with the emitter resistor, R_E. The capacitor acts to couple any changes at the emitter to circuit common. In a sense it

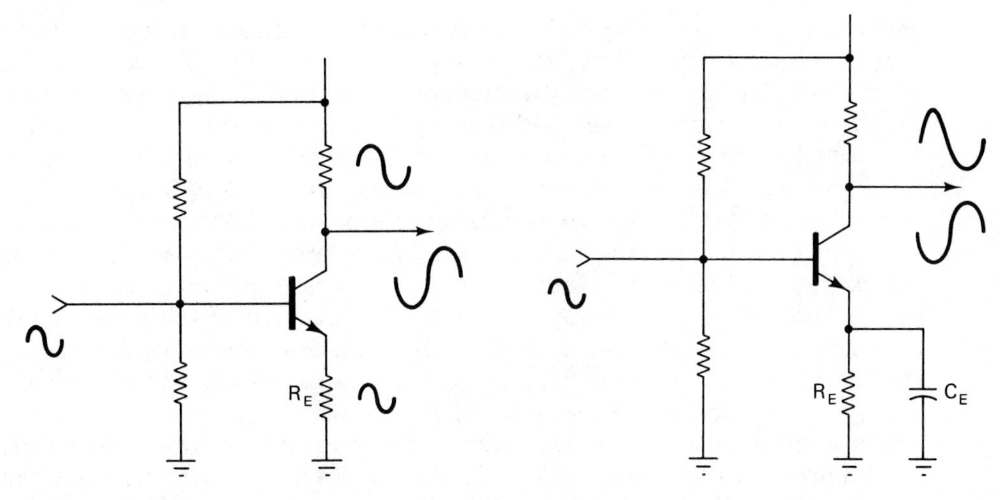

Figure 11-5 When the emitter resistor is in the circuit, a signal voltage drop develops across it and degenerates the signal.

Figure 11-6 A method of solving the degeneration problem is to use an emitter-resistor bypass capacitor.

bypasses the signal around the resistor to circuit common. The result is a constant dc voltage at the emitter, and little chance for thermal runaway to occur. The output signal is much larger because there is no signal regeneration across the emitter resistor.

Common-collector circuit. The second circuit in this group is called the common-collector or emitter-follower circuit. This circuit is shown in Figure 11-7. The major difference between this circuit and that of the common emitter is the placement of the load resistor, R_1. This resistor is now con-

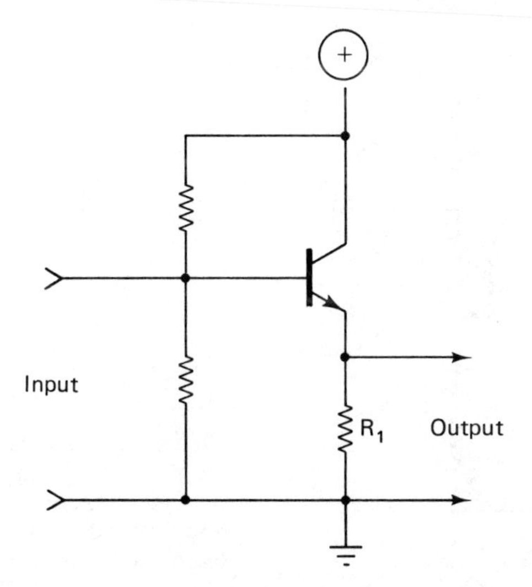

Figure 11-7 The common-collector circuit has its output connected across the emitter resistor.

nected between common and the emitter of the transistor. The collector lead is connected directly to the source positive terminal. All voltages that are used to develop the output signal are taken from the emitter load resistor. Its value is much larger than that used for thermal runaway protection. The signal voltage that develops across this resistor is almost as large as the input signal voltage. Its amplitude usually is about 98% of the input signal. The output voltage, or signal, is in phase with the input signal. A rise in conduction in the transistor now develops a larger, in-phase signal across the emitter load resistor. A drop in input signal level produces less conduction in the transistor. Resistor R_E is the static element in the series circuit, consisting of the load resistor, R_E, and the emitter–collector connections of the transistor. All resistive changes occur in the transistor. These result in a voltage change across the terminals of the transistor. The result is a change that also occurs across the load resistor. The output terminals of the circuit are connected to the load resistor. This produces a dc voltage variation that is in phase with the input signal and has the appearance of the input signal.

This circuit is used as an isolation or buffer amplifier. Its relatively low output voltage and high current output make it an excellent power amplifier stage.

Common-base circuit. The final amplifier circuit is called the common-base circuit. It is shown in Figure 11-8. This circuit uses both an emitter resistor and a collector resistor. The signal is injected across the emitter resistor. It causes the emitter voltage to rise. The effect is to produce less conduction through the transistor. The result raises the voltage developed across the collector-to-emitter junction. This produces a relatively high dc voltage at this point. Changes in amplitude and polarity of the input signal produce a high voltage that is in phase at the output terminals of the circuit.

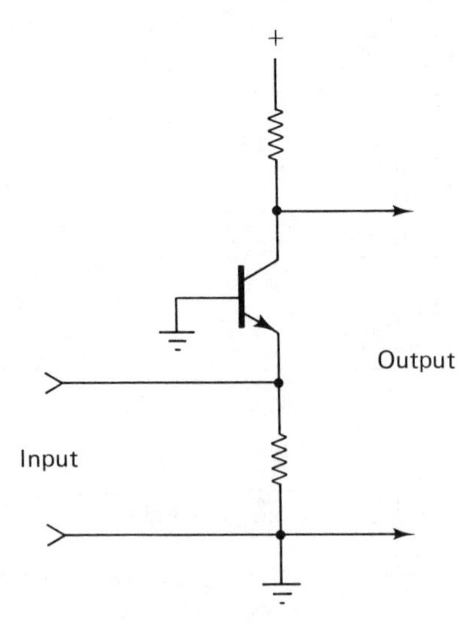

Figure 11-8 The common-base amplifier has its input connected between the emitter and common and its output between the collector and common.

This circuit has a very low input impedance. It also has a very high output impedance. Its main purpose is that of an input or front-end RF amplifier in a receiver. This circuit is often used as an RF amplifier in FM tuners and receivers.

Transistors as controlled resistors. The analysis of a transistor amplifier circuit becomes very simple when one considers that this device acts as a controlled variable resistance in the circuit. The output circuit consists of the emitter and collector terminals and some type of load resistance. These are series connected in order to obtain a voltage drop across each device. The transistor, acting as a variable resistance in the circuit, develops a varying dc voltage at the output terminals due to the action of the input signal voltage. This varying dc voltage at the output is identified as the "signal." This varying dc voltage is then coupled to the next stage of amplification in an audio system.

BIAS AND CLASSIFICATION OF THE AMPLIFIER

For practical purposes most audio amplifiers are classified as class A, AB, or B systems. This classification is totally independent of the common-emitter, common-base, or common-collector designation. This new classification is directly related to the duty cycle of the amplifier stage. A transistor amplifier can be designed so that it operates during all of the input signal cycle. This is called a *class A* amplifier. The same amplifier can be designed to operate on a 50% duty cycle. This is usually done when the input signal has either its positive or its negative half-cycle. This is called a *class B* amplifier. The amplifier can also be designed to operate for less than half of its duty cycle. This is called a *class C* amplifier and is usually used only in RF amplifiers in transmitting equipment.

The type of operation for any amplifier is dependent on how it is biased. When the base bias is established at the midpoint of an operational curve, the transistor is always on. An input signal causes the base bias to swing above and below the operating point. Current always flows in this circuit. The output wave is a faithful, but larger reproduction of the input signal. This type of operation is designated class A amplifier operation.

The bias point for the transistor can be located at any point of the operational curve of the transistor. When the bias point is located at the transistor's cutoff point, the transistor is turned on only when the signal moves the bias into the operating area of the transistor. This causes the transistor to operate as a diode. It conducts only during one-half of the input-wave duty cycle. This type of operation is called class B. Its main application is in push-pull power amplifiers.

Two transistors are employed in the push-pull output circuit. An ideal circuit design has one transistor turning off at precisely the same moment that the other transistor starts to conduct. In a practical circuit, this is very difficult to do. The bias point is moved so that the transistors are biased just before the point of cutoff. This arrangement permits a smooth transition for conduction. It also provides a faithful reproduction of the signal waveform. The classification for this is neither class A nor class B. Since it uses the better features of both classifications, it is called class AB. Most audio amplifiers use this classification for their output stage.

The third type of amplifier is called class C. This amplifier is biased well past the point of cutoff. It operates for less than half of the input-wave duty cycle. This amplifier uses a resonant *LC* circuit to provide a signal during the period that it is not conducting. The waveforms associated with these amplifiers are shown in Figure 11-9. The solid line in each drawing represents the conduction period for the amplifier. The dashed lines represent that part of the duty cycle when the amplifier is not conducting. The only classification that provides full-wave output is the class A system. This is the system that is used for most low-level small-signal amplifiers.

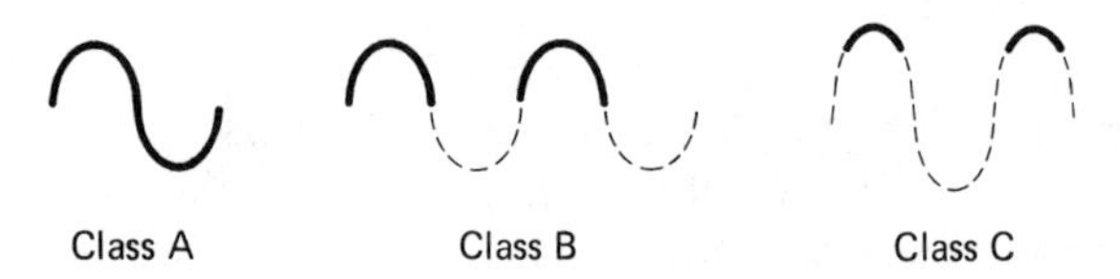

Figure 11-9 Classification of amplifier by their duty cycle produces the waveforms shown above.

SIGNAL TRANSFER SYSTEMS

There are very few single-stage amplifiers in use at the present time. Most audio amplifiers are made up of several stages of amplifiers. These amplifiers are designed so that the signal can be transferred from one stage to the next stage. There are three basic methods that are used to transfer signal between stages. These are identified as direct-coupled, *RC*-coupled, and transformer coupled. Each of these methods is described in this section. Methods of troubleshooting each system are also discussed.

Direct coupling. This method of coupling is often used in low-frequency audio amplifiers. It responds well to the transfer of the entire range of audio frequencies, with little or no distortion. A schematic drawing of one type of direct-coupled amplifier is shown in Figure 11-10. The significant point in this circuit is the direct connection between the collector of Q_1 and the base of Q_2. The bias for the first stage, Q_1, is developed from the voltage-divider network consisting of resistors R_1 and R_2. A signal injected at the base of transistor Q_1 will cause its collector voltage to change. It will rise as the input signal falls. In other words, as Q_1 is turned on by the input signal, the dc voltage at its collector decreases.

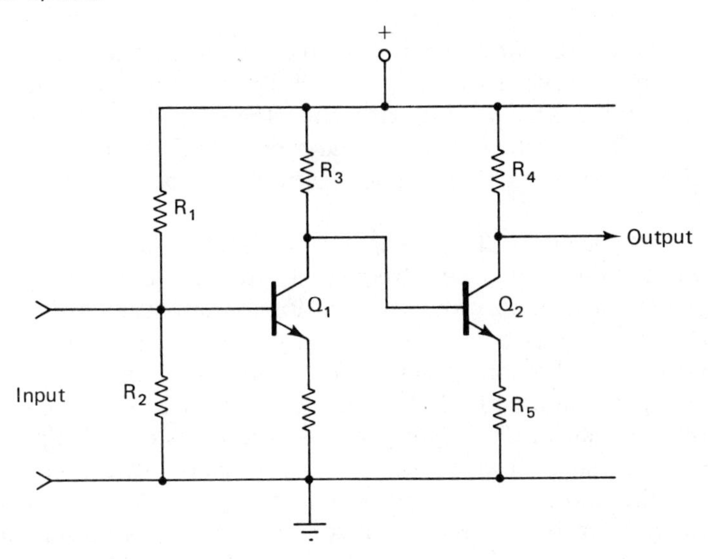

Figure 11-10 Direct-coupled amplifier circuit. Direct coupling is between Q_1 (collector) and Q_2 (base).

The varying dc collector voltage is applied directly to the base of the second transistor, Q_2. This dc voltage develops the bias for the second stage. The varying dc signal voltage is coupled from one stage to the next in this manner. A major consideration for a circuit of this type is the ability of the second transistor to operate in class A with a fairly high dc voltage applied to its base. In addition, the base signal swing as it goes from positive to negative is fairly large. The second transistor must be able to handle this swing without going into saturation or cutoff.

Troubleshooting the direct-coupled system. There are three faults that normally occur in any amplifier system. These are no signal output, distorted signal output, and low signal output. Let us examine each of these and how to troubleshoot it.

A no-signal output can be diagnosed as either an open or shorted component or a defective transistor. The direct-coupled amplifier is unique in that one stage can affect another stage. Signal tracing is not always the best method of analysis for this system. The use of a voltmeter and ohmmeter will often help to locate a problem area. The first test is to determine if the proper operating voltage is present. If this is good, measure the emitter-to-base voltage of the transistors. This test will show an open or shorted transistor very quickly. If the normal 0.7 V difference found in a silicon transistor is present, the base-emitter circuit is good. A comparison of transistor terminal voltages with those provided in the service literature will quickly identify a trouble area. Follow this with dc circuit analysis to locate a specific component that has failed.

When the problem is a distorted signal, the solution to the problem is found in a wrong bias voltage. A second reason for this type of problem is overdriving. This means that the stage has too much signal applied and is being driven into saturation and/or cutoff. Bias voltage is checked with a voltmeter. Overdriving can be observed with an oscilloscope.

RC coupling. This coupling method is used in circuits that operate over a wide range of low frequencies. It is used in place of the direct-coupled system because it provides isolation of operating voltages between stages. It is also much easier to localize a problem due to the stage isolation in the system. A schematic diagram for a two-stage *RC*-coupled audio amplifier is shown in Figure 11-11. In this circuit capacitors C_1, C_2, and C_3 are called coupling capacitors. Their purpose is to transfer or couple changes in voltage from one stage to the next stage. The major difference between this system and the direct-coupled system is that only the *changes* in dc operating voltages are transferred. Resistors R_1 and R_2 are used to establish the operating bias on transistor Q_1. Resistors R_5 and R_7 do the same for transistor Q_2. Resistor R_3 is the load resistor for transistor Q_1. Changes in the dc operating voltage that occur at the collector of this transistor are called the *signal*. These changes will develop a varying charge in the left-hand plate of capacitor C_2. These varying charges will cause the charges on the right-hand plate of capacitor C_2 to vary. When the voltage on the collector

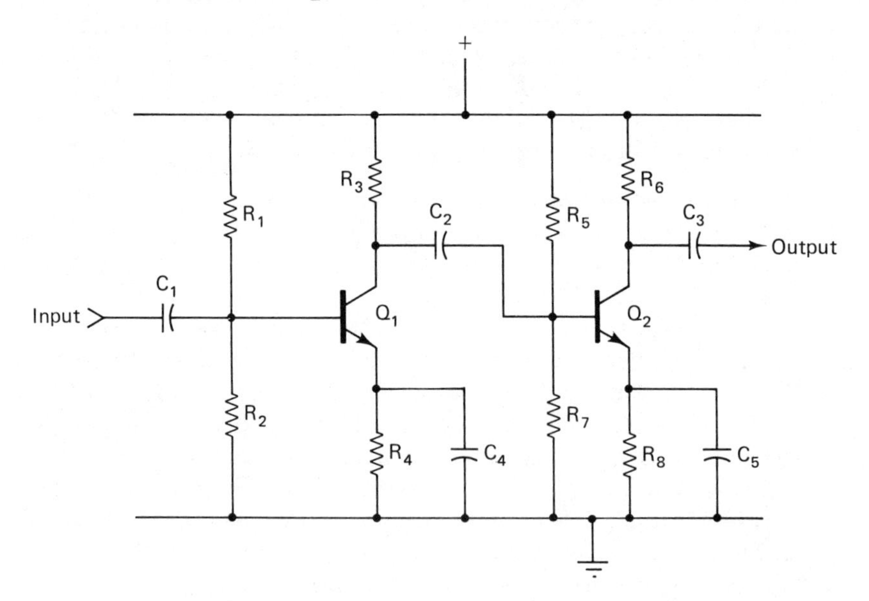

Figure 11-11 Capacitive coupling is used in this circuit. Capacitors C_1, and C_2, and C_3 are used to transfer signals from one stage to the next one.

side of the capacitor rises, the voltage on the base connection side of this capacitor will fall. The changing charges on the plates of the coupling capacitor can be observed on the screen of an oscilloscope. These changes are also called the signal. This is the manner in which the signal is transferred from one stage to another by means of a coupling capacitor.

This circuit is called an RC-coupling circuit because it requires a resistor in addition to the coupling capacitor. Resistors R_2 and R_7 are used in the circuit together with capacitors C_1 and C_2. These resistors provide an electron flow path for the changes that develop on the plates of the coupling capacitors.

Troubleshooting the RC-coupled system. This system is best tested with an oscilloscope. It is a linear-flow-path system. First check to see that the input wave is present. Then split the system in half by making the next check at capacitor C_2. If signal is present at this point, the trouble is after the test point. If signal is not present, the trouble is between the test point and the input. An oscilloscope test will quickly determine if the transistor is amplifying. After this test, use a voltmeter or an ohmmeter to test the dc current path of the transistor. A lack of any signal usually indicates a bad transistor or series resistor. Weak or low amplification is often caused by degeneration in the circuit. This is tested by bridging the emitter bypass capacitor C_4 with a capacitor known to be good. Distortion in the system is caused by a leaky transistor, incorrect base bias, or too high an input signal. These are checked by the usual voltage and resistance test methods.

Transformer-coupled system. The final method of transferring a signal from one stage to another uses a transformer for this purpose. A schematic diagram of a transformer-coupled amplifier is shown in Figure 11-12. Trans-

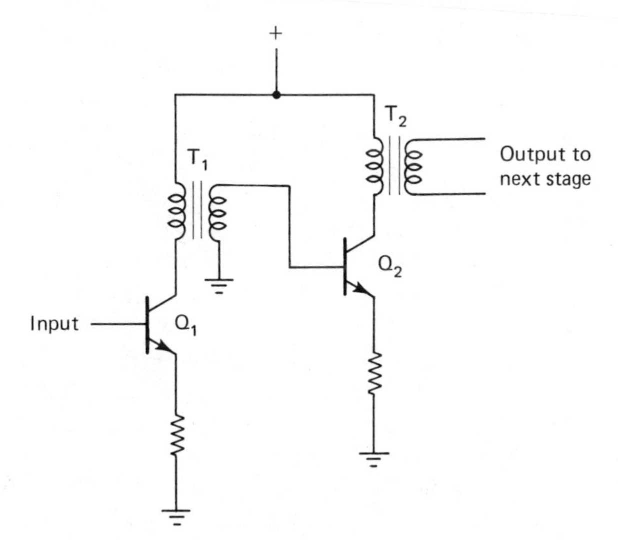

Figure 11-12 Transformer coupling is used in this circuit for signal transfer purposes.

formers T_1 and T_2 are all used as coupling devices between the stages of this amplifier. The primary of each transformer is connected in series with one of the transistor emitter-to-collector current paths. The transformer will develop a magnetic field in its primary winding as current flows through this winding. The magnetic field varies in strength depending on the amount of current flow. The current will vary as the amplitude of the input signal causes each transistor to change its interval resistance. This change in current flow changes the strength of the magnetic field and induces a voltage onto the secondary windings. The changes in voltage in the primary induce a voltage onto the secondary windings and transfer the signal from one stage to another in this manner.

Troubleshooting the transformer-coupled system. This system, like its *RC*-coupled counterpart, is best tested with an oscilloscope. It is also a linear-flow-path system and can be divided in half until the signal-loss area is located. The transformers provide isolation between stages. This makes troubleshooting much easier because the stages are electrically separate from each other. Once the defective stage is isolated, the normal voltage and resistance checks are made to locate a defective component. Typical problems in these stages include a complete loss of signal, distortion of the signal, and low gain in a stage. Each of these problems is resolved by using accepted troubleshooting techniques.

Integrated-circuit amplifier system. All the signal transferring systems just described are found in systems that utilize integrated circuits. The only major difference that applies to IC circuits is that one does not have the ability to change individual parts. The IC is an all-or-nothing type of device. When one section fails, the entire IC is replaced. Direct-coupled amplifier circuits are contained in the IC. Capacitors and transformers used for signal coupling are mounted on the circuit board next to the IC. Troubleshooting the IC amplifier is done with an oscilloscope. One looks for signal loss or distortion of the signal. If the supply voltage is correct and any other components are in working order, the IC is replaced. There is little choice except to replace it with one with exact specifications.

Troubleshooting, as mentioned earlier, is not very difficult. One has to know how the circuit is supposed to work. One must also know the purpose of each component. The correct use of the proper piece of test equipment saves hours of work in the shop. Develop a thinking attitude about each repair. Use a logical systematic approach to locating the problem area and the specific component. A few extra minutes learning this technique during the early days of repair will save hours of work later as you become more experienced. Let each set become a learning experience. After the repair is successfully completed, take a few minutes to analyze what you have done and think about the efficiency in locating the defect. Use this knowledge with later sets in order to become a highly competent individual.

QUESTIONS

11-1. What effect does the input signal have on the emitter–collector connections of a transistor?

11-2. What are the input, output, and common connections for the three basic amplifier circuits?

11-3. What is thermal runaway, and how is it reduced in an amplifier circuit?

11-4. What is the purpose of the emitter capacitor?

11-5. What are the bias classifications for amplifiers? Name them.

11-6. Explain direct-coupling signal transfer.

11-7. Explain capacitor-coupling signal transfer.

11-8. Explain transformer-coupling signal transfer.

11-9. What piece of test equipment is used to observe input and output signals?

11-10. How is an IC amplifier tested?

Chapter 12

Repair of AM and FM Tuners

The systems used for tuning, amplifying, and demodulating the AM and FM broadcast signals have many similarities. When one reviews the block diagrams for these two systems, the similarities are very obvious. These two systems are shown in Figure 12-1. Start at the output end of the system. Both have demodulators and IF amplifiers. The frequencies and the process of demodulation are different, but the function is the same. The next block is the mixer. Each system uses a mixer. The FM tuner usually has a separate mixer stage. Often, the AM mixer and local oscillator are combined into one stage called a *converter*. The low strength of the FM signal requires an RF amplifier stage. The AM signals are usually stronger and often this stage is omitted from the AM section of the tuner. The similarities in block diagrams are so great that this is often used as the basis for the initial stages of troubleshooting a tuner system.

The major differences between the two systems lie in the frequencies used in each. The FM tuner receives RF signals that are broadcast between 88 and 108 MHz. The IF frequency for the FM system is 10.7 MHz. The AM tuner receives signals that are broadcast between 540 kHz and 1.6 MHz. The IF frequency for an AM radio is usually 455 kHz. The exception to frequency is found in most automobile radios. The IF frequency for these radios is 262 kHz. The other major difference occurs in the method of demodulating the carrier. A FM signal is frequency modulated. The demodulator system must respond to frequency changes. The AM signal is amplitude modulated. The demodulator system in an AM tuner must respond to changes in the amplitude of the signal.

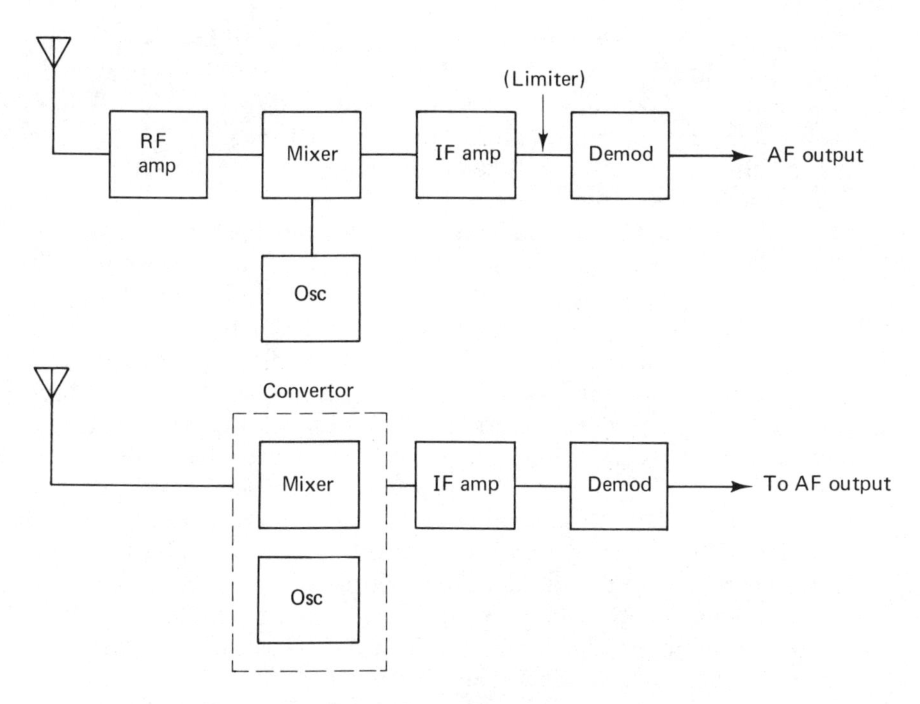

Figure 12-1 Block diagrams for radio tuners. Often, a converter stage is used instead of the separate amplifier, mixer, and oscillator blocks.

The final difference between these systems is shown in parentheses on the FM system block diagram. The amplitude of the FM signal is held to a specific design value for demodualtion. This requires some sort of amplitude-limiting process. It is called a *limiter* in the FM tuner. Often, the limiting action becomes a function of the demodulator. For this reason this is shown as an option on the block diagram.

The nice part about the diagnosis of radio tuners is that the similarities in the two systems can be used to simplify the location of a problem area. Procedures used for one system often apply also to the other system. It is advisable to learn the block diagrams. The basis for repair as explained in this book depends on a knowledge of how the system works.

GENERAL SYMPTOM DIAGNOSIS

The first major step in the repair of any system, including radio tuners, is to attempt to locate the fault area quickly. Often, this can be accomplished without removing the tuner from its cabinet. Let us look at how much can be accomplished in this manner. Start by turning on the set. If the dial lights or other lights glow, the set is receiving power. This eliminates one part of the power source system. The parts that are working include the line cord, any fuses in the primary part of the system, and the primary of the power transformer. The assumption at this point is that the tuner is connected to a speaker, and an amplifier and to an antenna system. The lamps should glow with what appears to be normal brilliance. A dim glow indicates some sort of overcurrent in the secondary of the power transformer. This is usually followed by smoke and a strong smell of overheated electronic parts. This, too, is an indicator of a problem area. Further investigation is required and it is necessary to make a visual inspection to locate the problem area.

Before the unit is removed from its case, there are other checks that can be done. One of these tests is to tune the radio from one end of the dial to the other. If the unit has a signal-strength meter, this can be used to check out certain blocks. If the meter moves upward at the location of known signals, the system is working from the antenna to wherever the meter is connected in the circuit. This is usually in the IF or demodulator sections. The problem is then located between the meter circuit connection and the output to the audio amplifiers. Another test that can be made is to use another signal source to test the local oscillator. If the problem is in an AM tuner, a second AM radio or tuner can be used to check out the local oscillator. Both tuners are turned on. The defective unit is tuned to a fairly high point on its dial, usually around 1.5 MHz. The second radio, or tuner, is placed as close to the oscillator section as possible. The dial on the nonworking radio is turned. If one hears a whistle or oscillation squeal from the working radio, its oscillator is working properly. Another way of doing this is to use a RF signal generator. Place the leads from the generator as close to the oscillator section as possible. Tune to the oscillator frequency of the radio. This is the sum of the tuned signal and the IF frequency. An AM tuner that has its dial set at 1 MHz will have its oscillator at a frequency of 1.455 MHz. If the tuner works and produces an audio signal in the speakers, the problem is located in the oscillator section. The signal generator acts as a substitute oscillator and provides the correct IF signal. This procedure also works for a FM tuner.

Another test for proper function and elimination of the working sections is best done with a battery-operated unit. It can be done with a tuner that has an internal power source, but it is more difficult to set up. This procedure monitors the current used by the tuner. One test is to turn the dial and watch the current flow in the system. If the current shows increases as the tuner selects a local station frequency, the set is working properly up to the demodulators. RF, oscillator, and IF action is occurring. Increases in current are normal as signals are processed in the system. A lack of any changes in current indicates a lack of tuning action. Further tests have to be made in such a situation.

One can test the IF system by injecting a modulated IF signal at the antenna connections of the tuner. If the audio is reproduced by the speakers, the tuner IF and demodualtor sections are functioning properly. A series of tests such as the ones described in this section will quickly determine the problem area. This will save much time and expedite the repair of the problem area. The major idea is to develop a system of rapid and purposeful tests. The results of these tests can be used to eliminate those sections of the receiver that are functioning properly. The balance of the blocks are those in which the problem is located. If you concentrate on these areas first, the problem is usually quickly located. It might be wise to make out a diagnosis sheet. One such sheet is shown in Figure 12-2. This will help to develop a pattern for the successful analysis of the tuner and the ultimate location of the trouble area.

Once the area of the problem is located, specific troubleshooting techniques are followed. The first thing to do is to remove the unit from its cabinet. A thorough visual inspection is next. Concentrate on those areas

RADIO SERVICE DIAGNOSIS

Date________________

Make ________________ Model ________________ Serial________________

Customer Complaint:

Technician Analysis:
- Physical:
  - Cabinet:
  - Knobs:
  - Line Cord:
  - Other:
- Operational:
  - Dial Lights:
  - AM
    - reception
    - tuning
  - FM
    - reception
    - tuning
    - stereo
      - right channel
      - left channel
      - indicator lamp
  - Audio
    - Level
    - Quality

Diagnosis: (suspected area of fault)

Figure 12-2 Sample diagnosis sheet used by the technician to help localize a problem area.

where the trouble is located. Look for such things as broken wires or components and discolored resistors or other parts. This visual inspection will help to locate the problem area. Do not overlook the battery holders. These often develop corrosion and result in a low output voltage for the system.

When the unit is removed from the case, there are some other types of tests that can be made. These require the use of electronic test equipment. There are two approaches that are usually used for locating a defective block in the system. One method is signal tracing. A signal generator is used to develop a modulated RF signal at the input. An oscilloscope is then used to trace the signal through the system. If one has the luxury of having an oscilloscope that has a high-frequency response, it can be connected directly to the test areas in the tuner. If the oscilloscope has a low-frequency response, a demodulator probe is used. This probe will show the modulation and not the carrier. If the modulation can be observed, the carrier has to be present and one can assume that the tuner signal processing to that point is correct.

Use the methods described in Chapter 11 to troubleshoot this system. It is basically a linear-flow-path system. Split it at or near the middle. This will quickly eliminate one-half of the system. The second check is at the midpoint of the area that is left to analyze. This procedure will quickly reduce the problem area to one stage.

An alternative method of troubleshooting is to inject a good signal at the midpoint of the system. In this case it would be in the IF amplifier stages. The generator is tuned to produce the required IF signal with a modulation of between 400 and 1 kHz. The generator is moved to the appropriate stages and adjusted for the correct signal frequency. One must recognize that the signal developed at the output of the tuner is weaker as the generator is brought closer to the output because there are fewer amplification stages left in the system. Either the signal tracing or the signal injection method can be successfully used to locate a problem area in the tuner.

PARTS LOCATION

When the service information is minimal, the technician has to use previous knowledge to locate specific areas for blocks in the tuner. One such system is shown in Figure 12-3. There is a schematic diagram available for this tuner, but it does not have a parts layout. The diagram shows the location of specific sections. The reasons for this identification are given in the following section.

FM RF amplifier. Leads for this high-frequency section must be kept short. The RF amplifier is located near the tuning capacitor. Inductors used for these frequencies are small. These are also shown here.

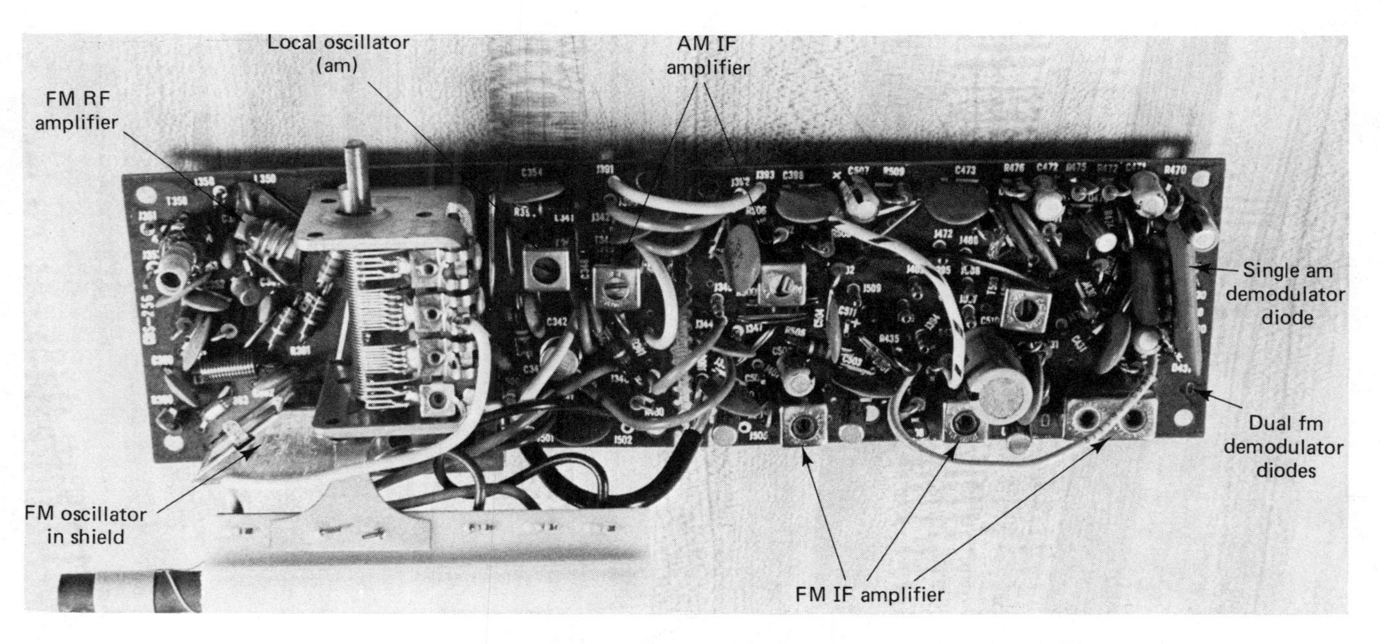

Figure 12-3 The use of a parts layout will often help identify sections of the set.

Local oscillators. These are also located near the tuning system. Oscillator coils are usually housed in metal shield cans. These are often identified by a color of paint on either the can or the tunable core. Unfortunately, the colors are not apparent in the picture.

IF amplifier. The FM tuner usually has more IF transformers than its AM counterpart. The IF transformers are usually close to each other. There are three transformers in the FM IF system compared to two used for the AM IF system.

Demodulators. The AM demodulator uses a single diode and the FM demodulator usually uses two diodes. The location of these diodes and the relation to the audio output connections will aid in the determination of the specific demodulator. Often, the IF and demodulator systems are located in a single line in the circuit board. The relationship of transistors, diodes, and IF transformers all aid in the ability to relate schematic drawings to exact sections of the tuner.

CIRCUIT DESCRIPTION AND REPAIR

The repair of any circuit is much simpler when one understands how the circuit functions. The major problem encountered by the service technician is that there is a lack of understanding as to how the circuit functions and what is supposed to happen in the stage or block. There seems to be a lack of ability to apply the knowledge about electronic circuits to the actual circuit. Once the technician overcomes this handicap, the circuits are quickly repaired. Repair techniques are very similar for most electronic circuits. These include analysis of signal flow paths and analysis of the current flow path. One must learn to use an oscilloscope and a multimeter properly in order to analyze a defective circuit quickly. The selection of the proper test equipment, knowledge of how to use it for maximum results, and an understanding of what the test results show will make you an excellent service person.

FM TUNERS

The input to the FM tuner is the RF amplifier stage. This stage has several important functions. These include a fairly large amount of gain, low internal noise, isolation, selectivity, and good linearity. The gain of this stage has to be high. It is usually on the order of 100 or so. This provides the amplification necessary for signal application to the mixer. The transistor has to have a low interval noise factor. If the noise factor is too high, this undesirable signal will be amplified as static throughout the entire tuner. Another important factor is isolation. The signals developed in the tuner oscillator cannot be allowed to feed back to the RF amplifier. If they do, these signals can be radiated and cause interference with other radios. The ability to select one station is also important. The resonant circuits used for selection have to have a bandwidth that will permit only the desired signal to be amplified by the system. The final characteristic is good linearity. This means that the gain of the stage is almost equal at all frequencies being received.

The input signal from the antenna or its feed line must be transferred with a minimum of loss to the RF amplifier. This requires an impedance-matching circuit in most circuits. A device called a *balun* transformer is often used for this purpose. The schematic diagram for a balun transformer is shown in Figure 12-4. The name "balun" is derived from the fact that it matches a balanced line to an unbalanced line. The 300-Ω antenna lead is not connected to circuit common. The term *balanced* is used to describe this circuit. The output of the circuit has one lead connected to circuit common. This is called an *unbalanced circuit*. These transformers are used to match the 300-Ω antenna lead to a low-impedance input circuit. This is necessary when a bipolar type of transistor is used. The circuit diagram for this type of circuit is shown in Figure 12-5. The input transistor is connected as a common-base amplifier and is used for RF amplification. This provides a high-voltage gain at the input. The parallel resonant circuit consisting of VC_{101} and L_{102} is used to tune or select the desired resonant frequency. The capacitor, being variable, is able to change the resonant point of the circuit.

A second circuit for the same purpose is shown in Figure 12-6. This diagram illustrates the circuit block diagram and the actual circuit. A field-effect transistor (FET), Q_1, is used as the FM RF amplifier. The selection of a FET provides a high input impedance and fewer losses when coupling the antenna to the RF amplifier. This also permits a higher input sensitivity rating for the tuner. This circuit has a resonant circuit at the input. It con-

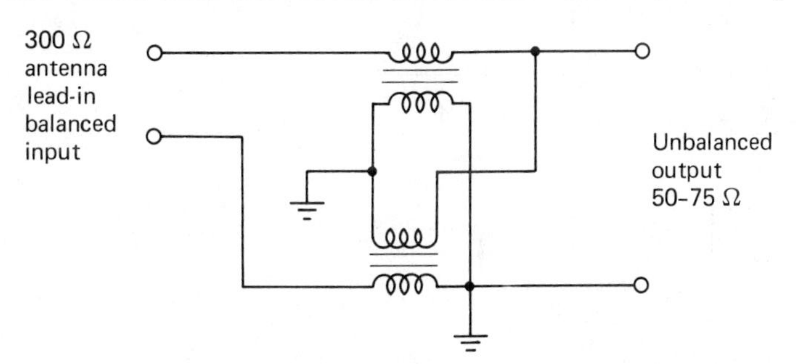

Figure 12-4 An impedance-matching transformer for antennas is called a balun.

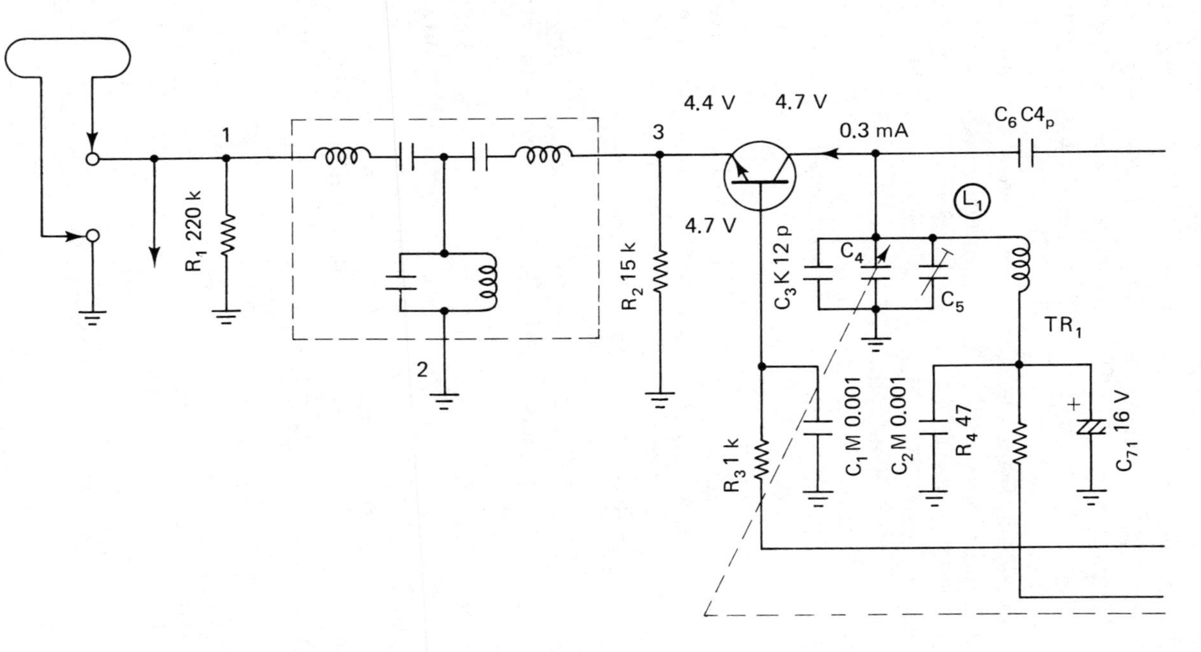

Figure 12-5 Many FM receivers use a common-base amplifier in the RF section because of its high gain factor.

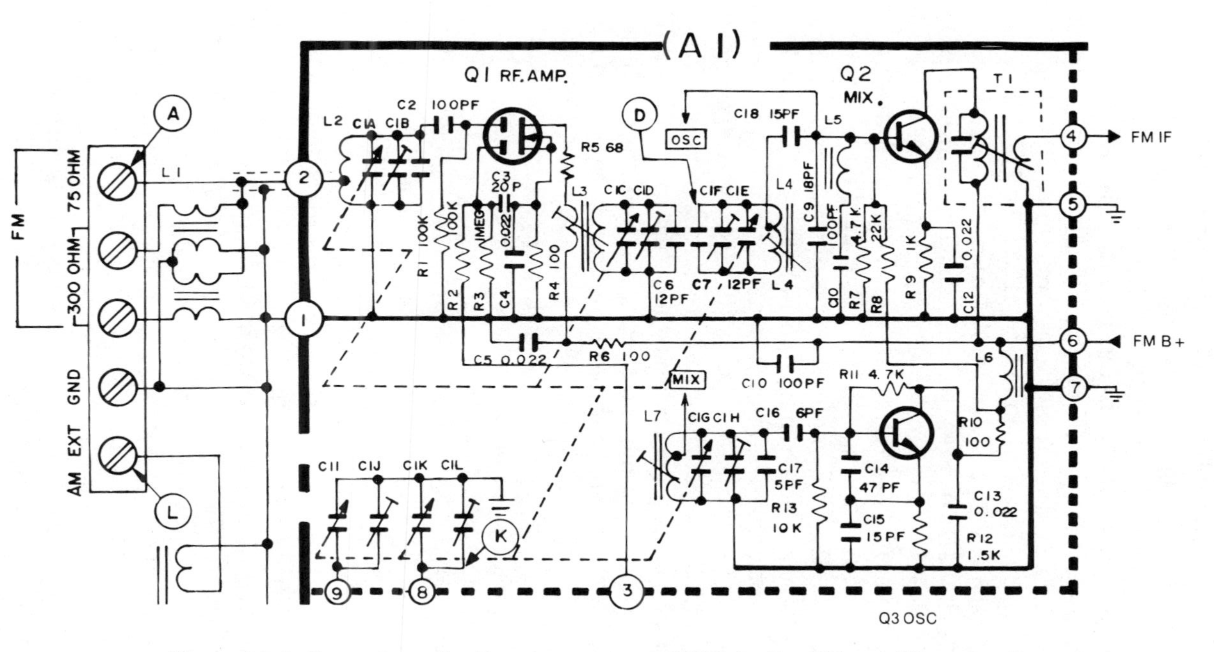

Figure 12-6 Current production tuners use a FET in the RF amplifier circuit.

sists of L_2 and a part of C_1. This is a parallel resonant circuit. It develops a high impedance and a high voltage at the point of resonance. The output circuit of the RF amplifier is tuned by L_3 and another part of C_1.

The RF amplifier output is coupled to the mixer stage. Its input is the resonant circuit C_{1F} and L_4. Induction coupling is used to transfer the signal between these stages. The oscillator circuit uses capacitor C_{1G} and inductor L_7 for its resonant circuit. Its output is connected to the base of the mixer transistor through resistor R_8. The output of the mixer is tuned to the 10.7-MHz IF frequency by transformer T_1. This transformer is tuned to the IF frequency. It has a bandwidth of at least 150 kHz in order to pass the FM stereo signal. A pair of resonant ceramic filters are used to limit the IF signal to its 150-kHz bandwidth. The signal is amplified between these stages by the IF amplifier transistor Q_{201}. The signal goes next to an IC IF amplifier, IC_{201}. This IC has several functional blocks. A diagram of the contents of this IC is shown in Figure 12-7. The modulated IF signal is processed through several stages of amplification in the IC.

Demodulation of the monaural FM IF signal is done by using a quadrature coil. It demodulates the IF signal by combining two versions of the signal that are 90° out of phase with each other. The resulting signals are integrated and the modulating information is removed. The audio signal is then fed to an audio amplifier block and then to the output of the IC.

This IC also has a muting circuit. The muting circuit is often called an audio squelch circuit. Its purpose is to change the bias on the audio amplifier. When an audio signal is present, the audio amplifier is biased so that it is on. When no signal, or too low a signal, is present, the mute circuit changes the bias on the audio amplifier so that it is off, or muted. It acts as an on–off switch for the audio block in an IC.

Another version of the IF amplifier section of the FM tuner is shown in Figure 12-8. This system is similar to those that use IC technology. Its difference is that it uses three transistors and IF transformers. Each of these transistors operates as a class A amplifier. The circuits are tuned to 10.7 MHz by the IF transformers. A space- and cost-saving feature in this circuit is the use of two IF transformers in series. One transformer in the series is used for FM and the other one for AM. The IF signal is passed only through the appropriate transformer. The other transformer acts as a piece of wire and for all intents and purposes is not even in the circuit.

The demodulation system used in this schematic is different from that used in the IC circuit. It is called a *ratio detector*. The secondary of the IF transformer T_4 is connected to two diodes and a resistive–capacitive circuit. The secondary on the transformer has a center-tapped winding. It is tuned to 10.7 MHz. Any positive frequency variations cause a current flow through D_8 and charges capacitor C_{39}. Negative frequency variations charge capacitor C_{40} in a similar manner. The voltages developed in these two capacitors are the same as the audio signal used to produce the frequency modula-

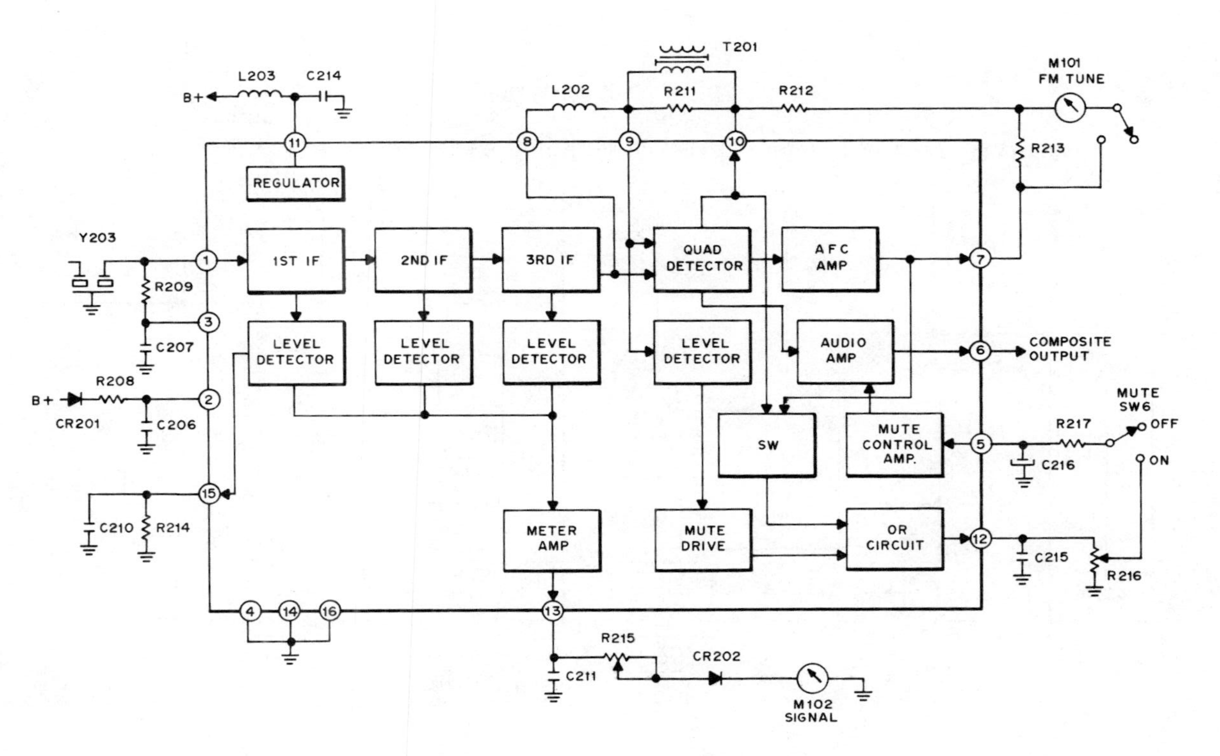

Figure 12-7 Block diagram of an IC used as an IF amplifier and demodulator. (Courtesy Zenith Radio Corporation.)

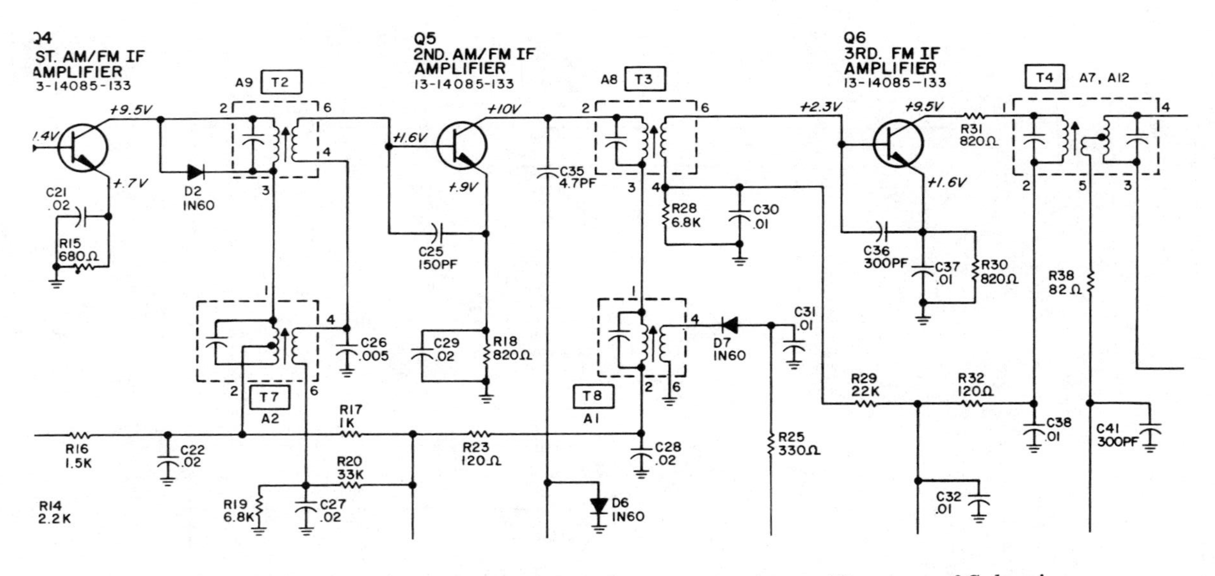

Figure 12-8 IF amplifier circuit using discrete transistors. (Courtesy of Sylvania Technical Services.)

tion at the transmitter. These variations, or signals, are coupled through an additional winding on the demodulator transformer and then to the audio amplifier.

Troubleshooting the FM tuner. The signal path system in the radio tuner uses both a meeting and a linear flow system. The processes listed at the beginning of this chapter should be used to locate the area of trouble. Signal injection is the most convenient method of localizing the trouble area if there is some doubt. Inject a modulated 10.7-MHz IF signal at the input to the IF amplifier. This will test the IF and demodulator sections. The signal can be injected next at the last IF. This will produce a very weak audio output due to the lack of sufficient amplifier stages between the test point and the audio output point. One problem that often occurs is that the capacitors in the IF transformers fail. This changes the point of resonance and does not permit passage of the IF signal. Unless the set has been moved around severely, there is little chance that the IF section requires alignment. Attempt this only when you have some experience. This is done as required and should always follow the procedure established by the manufacturer.

Transistor or IC failure is usually one of the most common problems. These are located by following the usual procedures for locating defective components.

AM TUNERS

AM radio tuners used in stereo equipment are often included by the manufacturer as a second thought. It is not possible to receive an AM stereo broadcast at this time. In addition, the limited frequency range of 0 to 5 kHz for the AM broadcast does not lend itself to full-range audio response. Most of these units are included to provide a more versatile stereo set. As a result of this, the cost of the tuner is kept to a minimum. Almost all AM tuners used in stereo systems are single-IC units. A schematic for one of these is shown in Figure 12-9.

The antenna for the AM tuner is wound on a ferrite rod. It is designated as L_{101} on the schematic. The converter block, which consists of a mixer and an oscillator, is found in the IC. Coil L_{102} is the oscillator coil and transformer T_{102} is the first IF transformer. This transformer contains a ceramic crystal IF filter and the usual resonant IC circuit. The second IF transformer is T_{101}. It is used to couple the IF signal between the two amplifier stages in the tuner. The audio output of the IC is connected from pin 12 of the IC. The demodulator is also included in the IC system.

Troubleshooting the AM tuner. AM tuner troubleshooting follows the suggestions made at the beginning of this chapter. Signal injection of IF signal

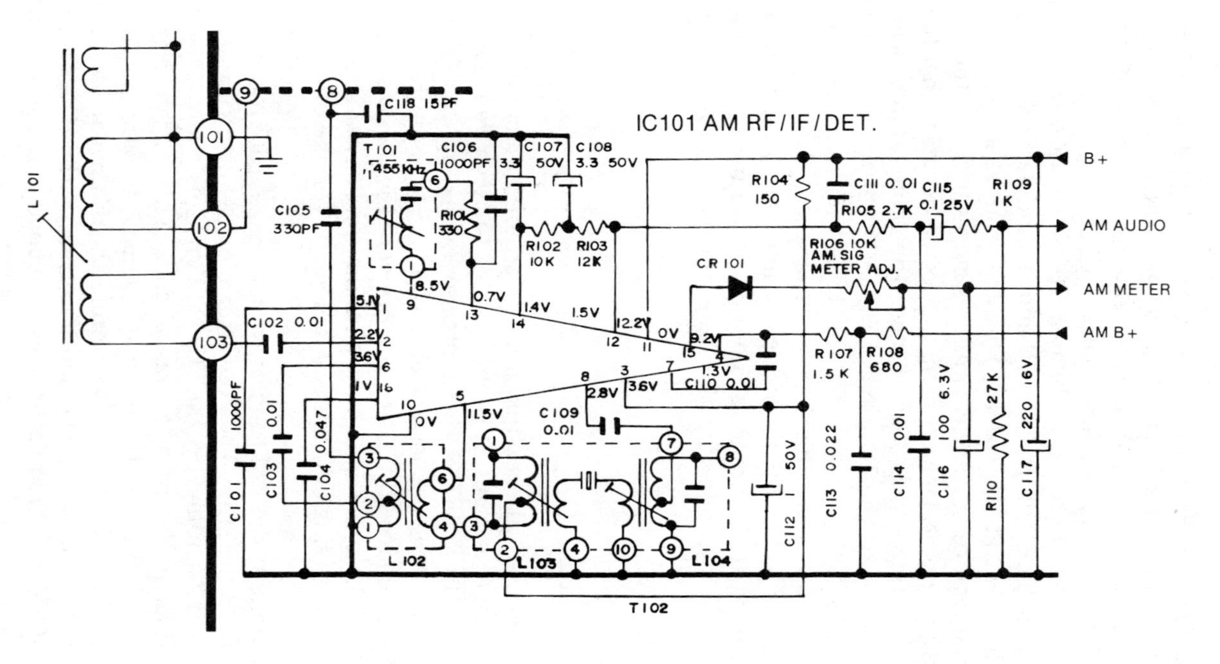

Figure 12-9 An IC used for AM radio reception has almost all components on one chip. (Courtesy Zenith Radio Corporation.)

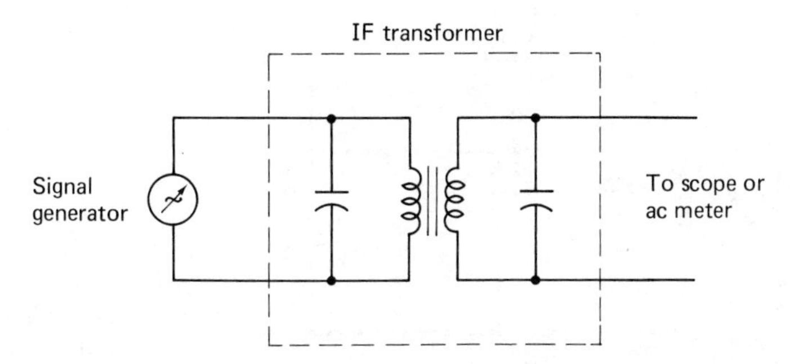

Figure 12-10 Method of measuring the frequency of an IF transformer using a signal generator and an output meter. At resonance the secondary voltage is very high.

will determine if that section and the demodulator are functioning. Additional tests should include measuring the voltages at the pins of the IC. If these are close to the values given in the schematic, the IC is probably good. Any voltages that differ greatly from the schematic values usually indicate a defective IC. Another test involves the IF transformers. These have a tendency to open. The open circuit could be either the winding or the resonant capacitor. The resonant frequency of an IF transformer is checked as shown in Figure 12-10. A signal generator is connected to one winding of the transformer. An oscilloscope or an ac voltmeter is connected to the other winding. The signal generator is then tuned through the operating range of the transformer. At the resonant frequency the output voltage is very high. At other than resonant frequencies the voltage developed in the secondary is very low. The resonant frequency of the transformer is determined by reading the frequency of the signal generator dial. This method is used to test IF transformers and to determine the frequency of an unknown transformer.

PHASE-LOCKED-LOOP (PLL) TUNERS

Many manufacturers of stereo systems in current production have done away with variable capacitive or inductive tuning systems. The present method of tuning a resonant circuit uses a special semiconductor called a *varactor diode*. This diode has the characteristic of being able to change the amount of capacitance between its elements. This is done by applying a dc voltage to the diode. A schematic for a FM RF amplifier that uses the varactor diode for resonating conductor L_1 is shown in Figure 12-11. The combination of this inductor and varactor diode D_1 are used to tune the input to the RF amplifier.

A second varactor diode, D_{21}, is used to tune the oscillator in this radio. This circuit is called a *voltage-controlled oscillator* (VCO). The fre-

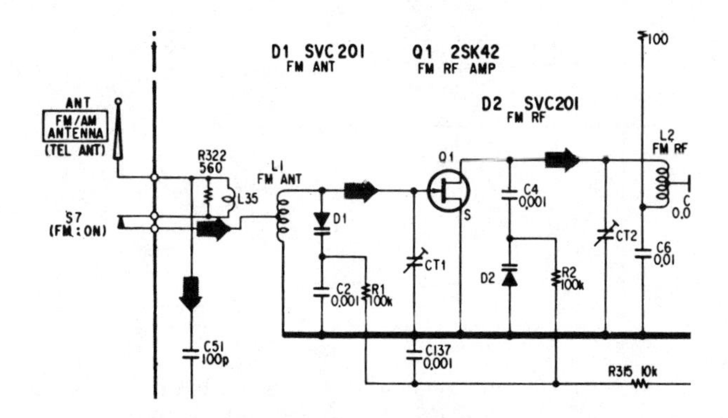

Figure 12-11 A varactor diode is used instead of a variable capacitor for tuning this circuit. (Courtesy Sony Corporation.)

quency of this oscillator is controlled by dc voltage. This voltage is developed by a phase-locked-loop circuit. The PLL circuit is programmed by either the dial or a keyboard-entry system. Changing the oscillator frequency will also change the frequency of the received signal. The dc control voltage that is developed in the PLL system is used to tune the varactor diodes in the RF amplifier and in the local oscillator.

Troubleshooting the PLL tuner. The complexity of this tuner system does not make it easy for the newcomer to electronic servicing. The only test that can be done is to measure the VCO control voltage. If this voltage does vary as the dial frequency is changed, the dc control voltage is working properly. If the oscillator frequency changes with the changes in dc control voltage, this circuit is also working. If the dc control voltage changes and the oscillator frequency does not change, the varactor diode is probably bad. When only one varactor is bad, it is wise to change all other varactor diodes in the control circuit. These are usually sold as matched sets. The major problem is the ability of the diodes to track together as they are tuned. This requires a great deal of experience in order to align the system properly. The procedure is best left to the experienced technician. The coverage in this book does not lend itself to a thorough discussion of digital electronic service. The author strongly suggests that the reader read one of the many excellent books written on this topic. Many of the newer sets use digital tuners and microprocessors for frequency control. This subject is one that should be explored by all newcomers and any others working in the field.

QUESTIONS

12-1. What frequencies are used in a FM tuner?

12-2. What section of the tuner is checked by heterodyning two signals?

12-3. What is the basic signal flow path for a radio tuner?

12-4. What blocks are located close to the tuning capacitor?
12-5. What parts are used to identify IF amplifiers?
12-6. What frequency signal is used to check IF amplifiers?
12-7. How is the varactor diode tuner checked?
12-8. How can one test the resonant frequency of an IF transformer?
12-9. What is the purpose of alignment?
12-10. How does a squelch or muting circuit work?

Chapter 13

FM Stereo Decoders

The ruling of the U.S. government that FM stereo broadcasting had to be compatible with FM monaural broadcasting led to the development of the multiplexed FM stereo system in use today. This system uses a combination of FM and AM signals in order to broadcast all of the FM stereo signal. The monaural portion of the signal is broadcast in an FM format. The stereo portion and the 19-kHz pilot reference signal are broadcast in an AM mode. This allows the stereo component to pass through the FM demodulator without modification. A block diagram of the FM stereo transmitter and the resulting stereo component signal is shown in Figure 13-1. The FM portion of the signal is developed in a normal fashion and is transmitted as such. The FM stereo component is added by multiplexing the AM signal on to the FM carrier. The resulting AM subcarrier is shown in the bottom of the figure.

The spatial requirements for a FM stereo signal are 150 kHz. This is 75 kHz on either side of the carrier frequency. A spectrum analysis of this signal is shown in Figure 13-2. The first 15 kHz of the signal is used for the monaural signal. This is followed by a low-amplitude "pilot" or reference signal at 19 kHz. The stereo component is multiplexed as a double-sideband suppressed carrier between the frequencies of 23 and 53 kHz. The frequency of the suppressed carrier is 38 kHz. The monaural component is called the L + R signal and the stereo component is called the L - R signal. The pilot signal is "clean" because there are no other signals close to its frequency to cause interference to it. The purpose of the pilot signal is to be a reference for the 38-kHz subcarrier regenerator in the decoder.

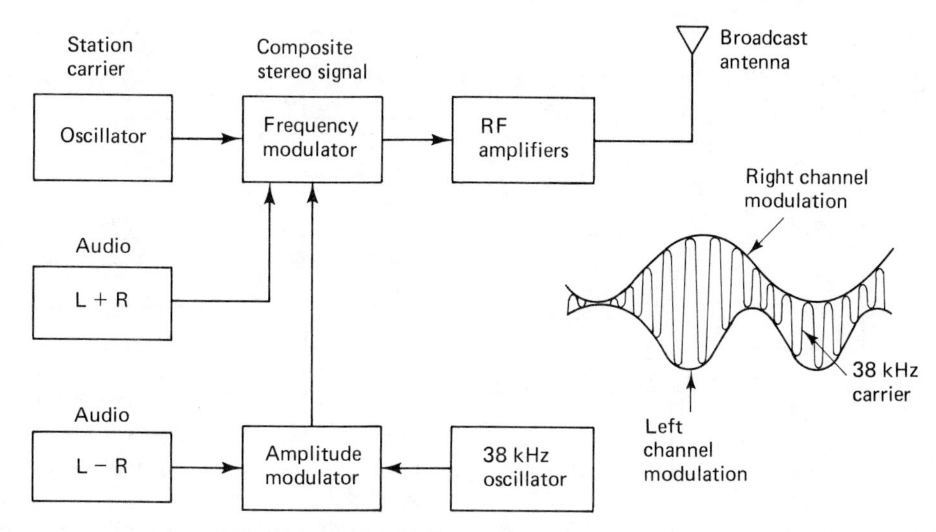

Figure 13-1 Block diagram of a FM stereo transmitter showing monaural, pilot, and stereo generation.

There are two popular methods used to reconstruct the stereo signal. One of these is called the *matrix* method. The second one is called the *switching* method. There are many matrix decoders in use. These are found in decoders that use discrete components and in IC decoders. The switching system is also found in both the discrete component systems and IC systems. The method that has become almost a standard today is the IC switching system. Each of these systems is discussed in this chapter. The theory of operation is reviewed and methods of troubleshooting are presented.

MATRIX DECODERS

The matrix decoder uses an adding procedure in order to reconstruct the left- and right-channel information. A block diagram for this system is shown in Figure 13-3. This is the same system as that discussed in Chapter 6. The

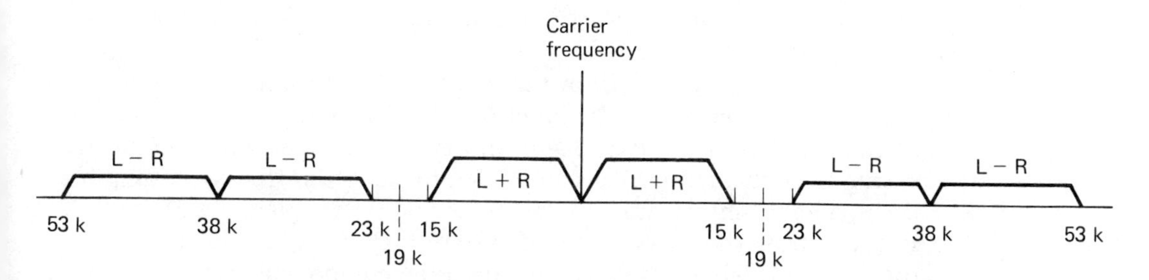

Figure 13-2 Spectrum analysis of the FM stereo signal, showing both upper and lower sidebands of the signal.

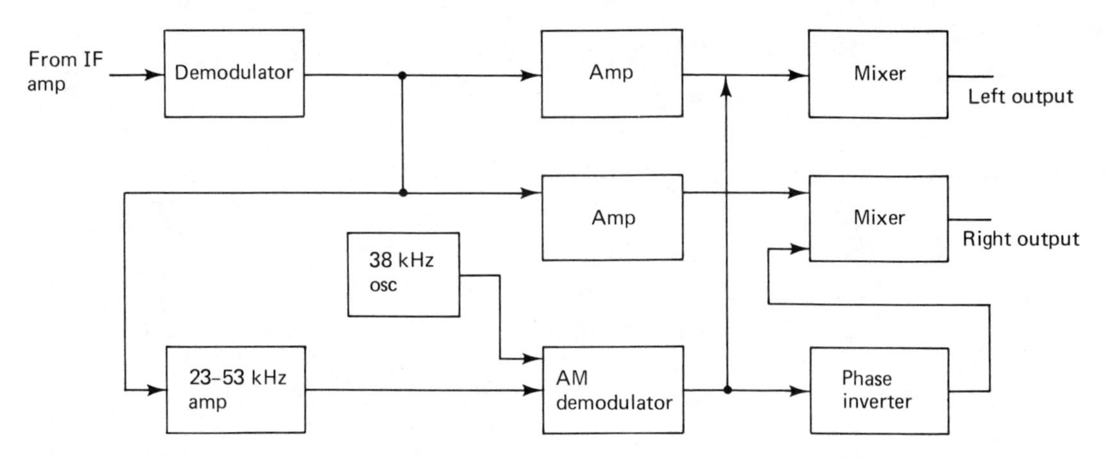

Figure 13-3 Block diagram of a decoder system used for FM stereo signals.

block diagram for this system and signal processing details are thoroughly covered in that chapter. The same system is shown in schematic form in Figure 13-4. The composite L - R, L + R, and 19-kHz tone signal is injected at the base of transistor X_1. The signal is amplified and sent to two different circuits. One of these is a 19-kHz amplifier. Transformers L_1 and L_2 pass this frequency to the 19-kHz amplifier, X_2. This signal is used to synchronize the 19-kHz oscillator X_3. The output of the oscillator is frequency doubled to 38 kHz by inductor L_3 and capacitor C_{14}. It is then amplified by transistor X_4 and sent to another amplifier. This circuit, X_5, is also used as a buffer or isolation amplifier for the stereo components. We will use a time-stopping explanation in order to follow the other signal paths in this system.

The composite L - R and L + R signals are fed from the emitter of X_1 to two other circuits. One of these is connected by a bandpass filter to the base of transistor X_5. At the base of this transistor the L - R and 38-KHz carrier are joined. The output of the transistor, at the emitter, is connected to two demodulators. One of these develops a positive-going signal. This is through diode M_1. The other develops a negative-going signal through diode M_2. The two diodes are connected to a matrix circuit. This circuit includes resistors R_{26}, R_{27}, R_{28}, and R_{29}.

The third signal path is from the emitter of transistor X_1. The signal is coupled from this element to the junction of resistors R_{27} and R_{28}. This is a part of the matrix system. The L - R and the L + R signals are combined in the matrix. Left-channel information is fed to the audio frequency amplifier X_6. Right-channel information is fed to the audio frequency amplifier X_7. The outputs from these two amplifiers are connected to the main audio amplifier system. This describes the discrete component matrix type of stereo decoder.

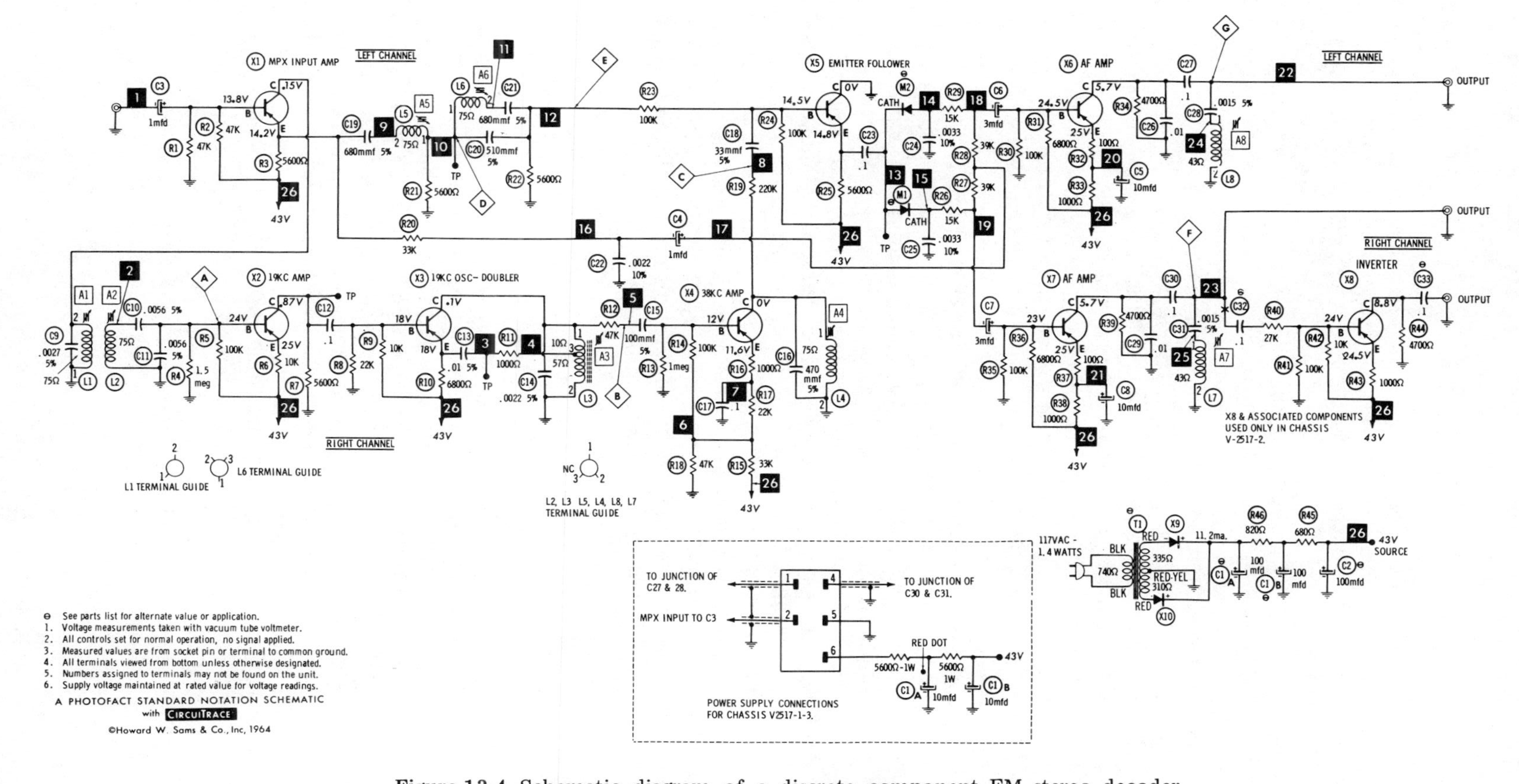

Figure 13-4 Schematic diagram of a discrete component FM stereo decoder. (Courtesy H. W. Sams Company.)

A later version of this system used an IC to accomplish the same thing. This is a much smaller unit, as the IC measures about $\frac{1}{4}$ by 1 in. All the electronics except capacitors and inductors are contained in the IC. This reduces cost, weight, and size of the stereo decoder. A schematic diagram for the circuits in the IC is shown in Figure 13-5. Included with the schematic diagram is a block diagram for the IC. Actually, the block diagram is of much more value to the serviceman. It is impossible for the average technician to repair a defective IC. The only repair is to replace it with a new one. The blocks are related to pinouts, or terminals, on the IC. The diagnosis of the IC becomes a signal-checking process. This is the same procedure that is used for the testing of the discrete component decoder. The difference, of course, is what happens after the tests are completed. Once the technician overcomes the fear of working with ICs, the discovery of how simple repairs can be is made. When the problem is found to be in the IC, the IC is replaced. This, except for testing, is the end of the repair.

When a discrete component decoder is serviced, the technician proceeds as with any other signal-flow-path system. This system contains separating, linear, and meeting signal paths. The method of repair depends on what defects are identified. If the monaural signal is present but there is a lack of stereo signal, both the L - R path and the 19-kHz and 38-kHz signal paths are evaluated. The area of the problem is reduced by making meaningful tests and evaluating the results of the tests. The schematic service employed for this circuit has identified test points in the system. These are used to measure for proper voltage or signal. The information derived from the test tells the technician how to proceed. Circuit analysis of the nonfunctioning circuit is done by following the current flow path of the circuit. This will lead to the location of the defective component in the circuit.

SWITCHING DECODERS

The decoder system that has become popular in the last few years is called a *switching decoder*. A block diagram for this system is shown in Figure 13-6. An electronic switch is used to switch the inputs to the FM transmitter. This switch operates at a very high frequency of 38 kHz. This is well above the audio-frequency range and rapid enough so that the listener is not aware of the switching process. The stereo decoder in the receiver also switches at a rate of 38 kHz. The two switches are synchronized by the pilot carrier's 19-kHz signal.

A schematic diagram for a switching stereo decoder is shown in Figure 13-7. The composite audio signal is fed to the decoder from the FM demodulator. It is amplified by transistor Q_1. This transistor has two outputs. One, from the emitter, is the composite stereo signal containing both L + R and L - R information. This signal is fed to the stereo demodulator trans-

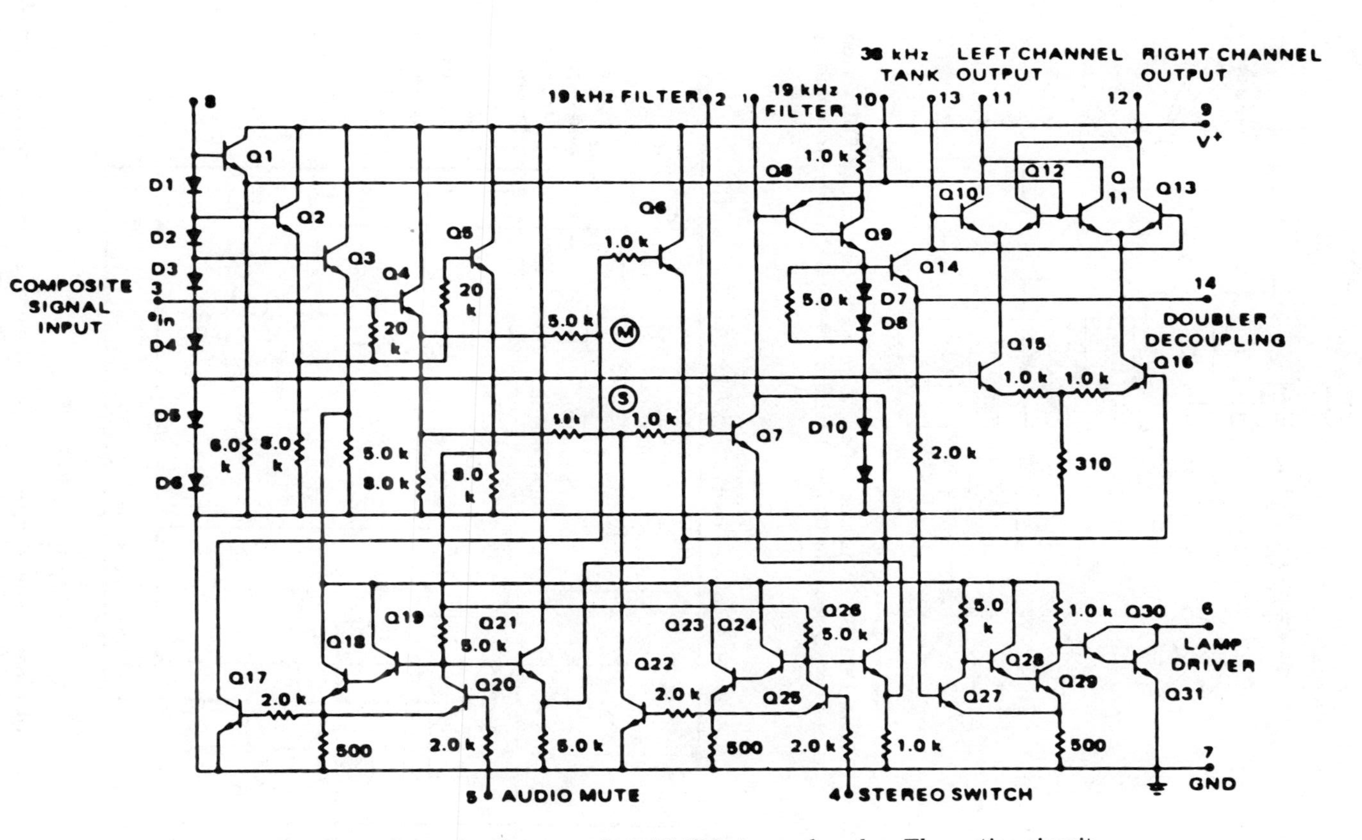

Figure 13-5 Schematic diagram of an IC FM stereo decoder. The entire circuit is contained on one IC.

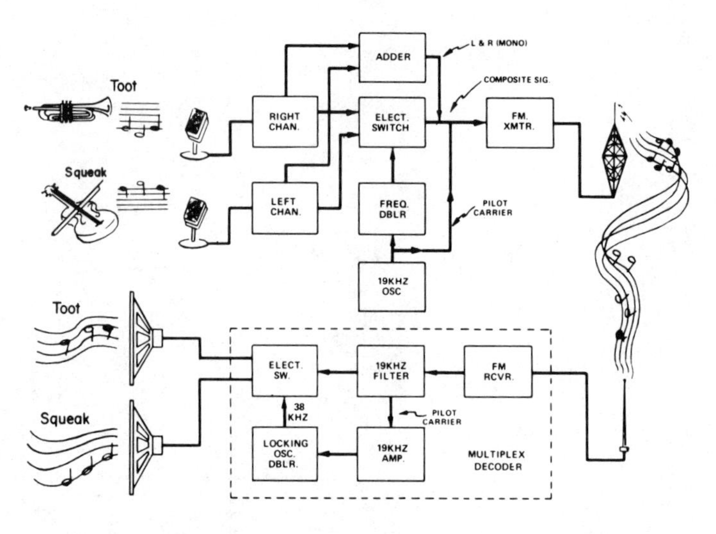

Figure 13-6 Block diagram of a FM stereo decoder using a PLL system. (Courtesy Delco Electronics Division, General Motors Corporation.)

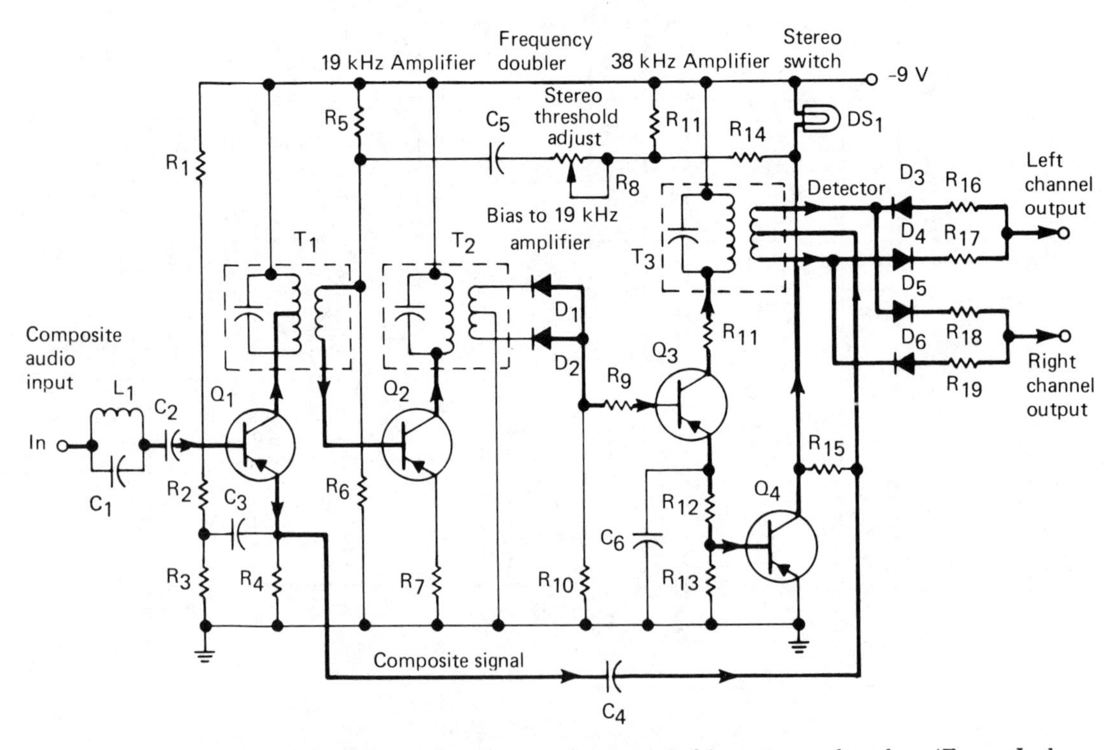

Figure 13-7 Schematic diagram for a switching stereo decoder. (From Joel Goldberg, *Radio, Television, and Sound System Repair: An Introduction*, © 1978, page 259. Reprinted by permission of Prentice-Hall, Inc.)

former T_3. Using our ability to stop time, we will leave the composite signal at the secondary of this transformer and trace the paths of the other stereo signals. The output from the collector of transistor Q_1 is the 19-kHz pilot signal. This signal is amplified by Q_2 and then fed to transformer T_2. The secondary of this transformer has a full-wave center-tapped output. It acts exactly as it does in a power supply. It doubles the frequency of operation. The 19-kHz signal is frequency doubled to 38 kHz in this manner. The output of this frequency-doubler circuit is amplified by transistor Q_3 and fed to the demodulator transformer T_3. Here it meets the 38-kHz switching signal for demodulation purposes. The output of this transformer is a bridge circuit that contains four diodes. The signal is switched to either the left-channel output circuit or the right-channel output circuit by the 38-kHz signal. The outputs from the bridge are the left- and right-channel audio signals.

The action described in the preceding paragraph can also be accomplished by using an IC. The circuit for this system and for the IC is shown in Figure 13-8. It is easy to understand this system when one returns to the block diagram that is used to develop the IC. It is the same block diagram that is used for the discrete component unit. The technician must know the block diagram for the system if it is to be serviced properly. The composite signal shown in Figure 13-9 is observed at the input to the IC. The 19-kHz and 38-kHz signals can be observed at the IC terminals that are connected to the two resonant circuits. These are pins 2 and 1. The audio signals that are developed in the IC are measured at pins 11 and 12. The other signal that is an output from the IC is the dc voltage used to turn on the stereo indicator lamp. This voltage is developed from a transistor switch in the IC. The 19-kHz pilot signal is used to bias the transistor to the "on" condition. A schematic diagram of this circuit is shown in Figure 13-10. The transistor is biased with no signal input so that it is in cutoff. This develops a high voltage at the collector. The voltage is close to that of the source. When the 19-kHz signal is present, the transistor is biased so that it is in saturation. This is an "on" condition. The emitter-to-collector voltage is very low under these conditions. This develops a large voltage drop across the indicator lamp and it turns on. The presence of a high and a low voltage at this point shows that the 19-kHz tone is being processed by the IC. If the voltage changes at the IC pin and the indicator lamp fails to light, the lamp is bad.

A modification of the switching system for decoding the stereo signal uses a different oscillator frequency and a phase-locked loop for holding the signal on the proper frequency. The block diagram for this system is shown in Figure 13-11. This system has more blocks than any of the other systems. It differs in that it uses a slightly different process to develop the decoded stereo signal.

Let us again use the luxury of stopping time to explain how this system works. The problem with all phase-locked-loop systems is that they do not

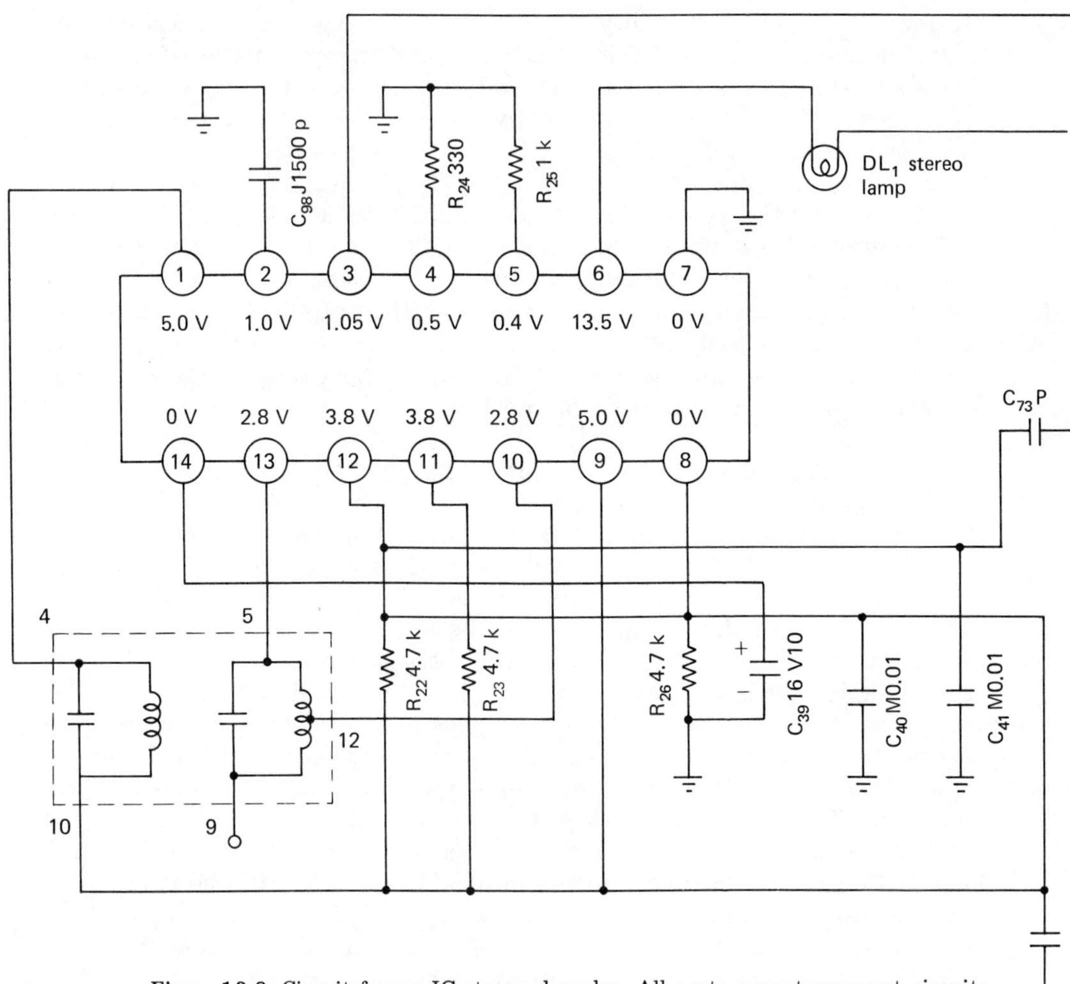

Figure 13-8 Circuit for an IC stereo decoder. All parts except resonant circuits are contained on the chip.

have a readily identifiable starting point. Let us start by injecting the composite stereo signal into the amplifier. The block diagram is convenient because the signal paths are identified for us. The signal, leaving the composite amplifier, takes these paths. Start with the pilot signal and follow its path. It goes to a 19-kHz pilot-phase detector block and then to an error amplifier. While this is happening the voltage-controlled oscillator (VCO) is developing a signal that is close to 76 kHz. The output of the VCO is fed to a divide-by-2 block, where it is changed to a 38-kHz signal. One output from the divide-by-2 block is connected to a second divide-by-2 block to reduce the signal further, to 19 kHz. This is then fed back to the pilot-phase detector block. In this block the two pilot signals are compared. The output

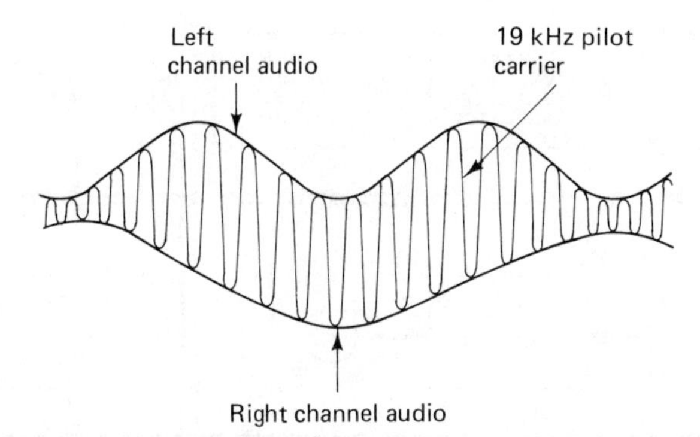

Figure 13-9 Waveforms usually found under test conditions at specific IC decoder terminals.

of the phase comparator is fed to the VCO. When both 19-kHz signals are in phase, the error voltage locks the VCO on 76 kHz. Any change in VCO frequency is corrected by the error amplifier. The dc voltage output from the error amplifier is used to keep the VCO on frequency.

The output from a second divide-by-2 block is used to operate the stereo indicator lamp. This 19-kHz signal is changed into a dc bias voltage as it is processed by a low-pass filter block. This bias voltage is used to operate the stereo indicator lamp amplifier switch block.

The last output from the first divide-by-2 block (76 to 38 kHz) is to the demodulator block. In this block the L - R signal is demodulated and combined with the L + R signal. The result is the separated left- and right-channel audio information. A pair of deemphasis circuits is used to filter out any remains of either the 19-kHz or the 38-kHz signals. From these blocks the audio signals are fed to the audio amplifiers.

Diagnosis of this system is very similar to that used for the non-PLL switching system. The first test is made with an oscilloscope to be sure that there is a signal entering the system. This is done at pin 2 or the IC. The next test is at the output terminals, pins 4 and 5. When a signal is present at these pins there is nothing wrong with the system. One could also check to

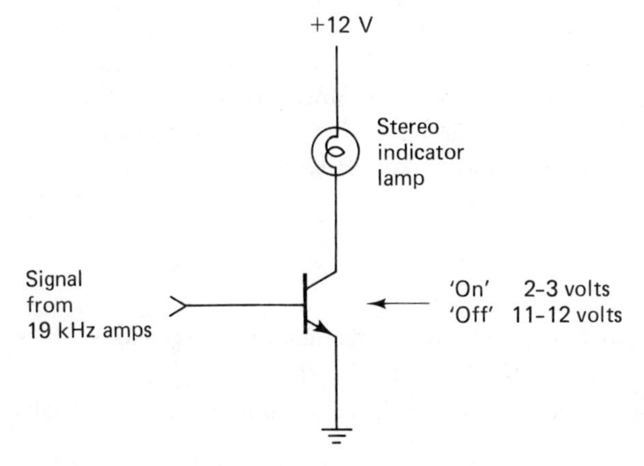

Figure 13-10 Schematic diagram for a transistor switch used to turn on a stereo indicator lamp.

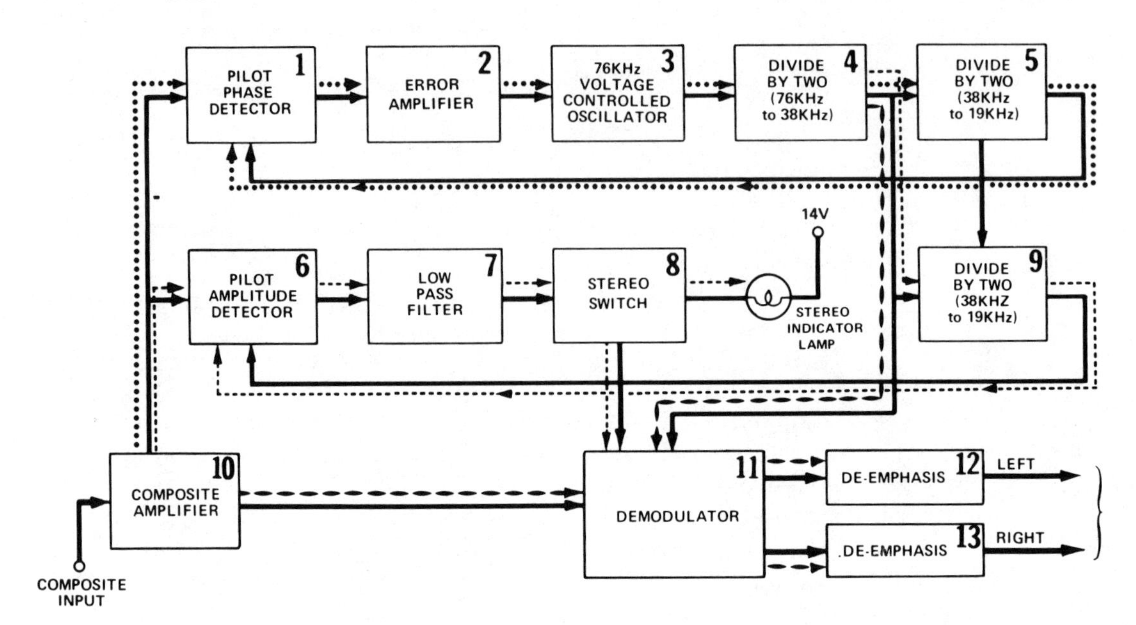

Figure 13-11 Block diagram for a PLL stereo decoder. (Courtesy Delco Electronics Division, General Motors Corporation.)

see if there is a square wave present at the 19-kHz test point, pin 12. This should also be checked with a frequency counter. Adjustment of the VCO frequency is done by a variable resistor. Adjusting the VCO also changes the divide-by-2 blocks. When the output at the 19-kHz test point is correct, the 76-kHz VCO is on the proper frequency.

It is possible to have everything working except the indicator lamp. When this circuit fails, it does not affect the other circuits in the IC. Unfortunately, if the repair requires that the lamp work, the entire IC must be replaced.

Analysis of any circuit really depends on an understanding of how the circuit functions. Use of the proper pieces of test equipment is important to the technician as the unit is serviced. In addition, the technician needs to know where to find both service literature and proper test points. The knowledge of how electron current flows and how basic circuits work will be all that is required for diagnosis and repair of the system.

QUESTIONS

13-1. Describe the signals found on a FM stereo carrier and their frequencies.

13-2. What is meant by the term "matrix" in electronics?

13-3. What part of the stereo signal is the L + R signal? What part is the L - R signal?

13-4. What is the frequency used for a switching decoder?

13-5. What frequency is used for the VCO in a switching decoder?

13-6. Describe how the stereo lamp circuit works.

13-7. Why is a "pilot" signal used on the FM carrier?

13-8. What kind of repair is done to a defective IC?

13-9. What is meant by the term "pinouts"?

13-10. What is meant by the term "error voltage"?

Chapter 14

Tape Decks

Recording and reproduction of sound by means of tape recorders is a major influence in today's stereo scene. The effect of a compact, self-contained tape system has gone a long way toward making tape units so popular. There are three systems in use today: cassette, cartridge, and reel-to-reel systems. The *cassette* and *cartridge* do essentially the same things. Each has a self-contained length of recording tape. Both are housed in a container. The major difference is in the type of container and in the method of moving the tape in its container. The third system, called *reel to reel*, uses two reels for the tape. One reel is full of tape and is called the *supply reel.* The second reel is empty and it is called the *take-up reel.* The tape moves from the supply reel to the take-up reel as it is recorded or played back through the system. These three major systems are shown in Figure 14-1. In essence, all three systems have a supply reel and a take-up reel. The cassette and reel-to-reel systems are easy to identify because both reels are separate. The cartridge system uses an endless tape and only one reel. The supply is taken from the inner part, or hub, of the reel and the take-up is at the outer part of the reel.

The systems used for moving the tape are but one part of the total tape system. This part is called the *transport*. Its purpose is to move the tape at a given speed past the record or playback heads. A second major portion of the system is the electronics section. A tape system may have individual amplifiers for recording and for playback. Most systems, however, use the same amplifier for both recording and for playback. When one amplifier is used, the input and output are switched between these two functions.

Figure 14-1 Tape systems usually found in stereo equipment include cassette, cartridge, and reel to reel.

There is one major difference between the systems used for playback and the system used for recording. This difference appears on the block diagram of the unit. One extra block is found in the record mode. This block, as shown in the diagram in Figure 14-2, is the *bias oscillator*. The bias oscillator is used only during recording. The three sections of the switch shown in the diagram illustrate this point. The switches are shown in the playback position. The input transducer to the amplifier is the tape head. The output from the amplifier is to the main audio amplifier and speakers in the system.

During the record mode the three switches are in their other position. The input device in the record mode is the microphone or some other device. The output from the tape unit is the tape head. In addition to these connections, the bias oscillator is turned on. The switch provides operating power to the block. There are two outputs from the bias oscillator. One goes directly to the erase head. The second output goes to the recording amplifier in order to reduce audio distortion due to magnetic fields in the head.

One additional block is often found in better-quality tape units. This block is called a *noise reduction system*. The system most often used for this purpose is called *Dolby* B* noise reduction. Its place in the block diagram

*Dolby is a trademark of Dolby Laboratories.

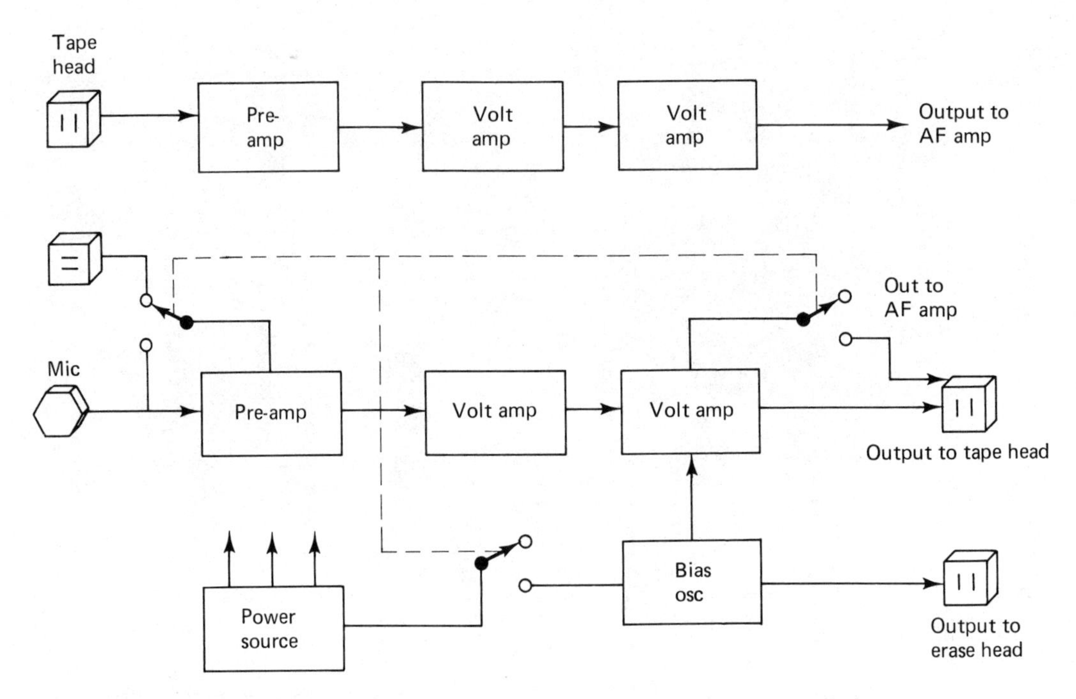

Figure 14-2 Block diagram for a record/playback tape deck system.

is shown in Figure 14-3. This noise reduction system is used to reduce unwanted tape noise, which sounds like a "hiss." A further explanation of this system is given later in this chapter.

One often hears the term "deck" used to describe parts of the tape systems. The *deck* is simply the tape player system without the capability to amplify the signals so that they can operate speakers. The tape deck consists of the mechanical transport, the electronic low-level amplifier, the bias oscillator, and any electronic noise reduction system. The systems described in this chapter are dual-channel and dual-purpose amplifiers. Only one channel for the system is shown on most of the figures, since the second channel is a duplicate of the first one.

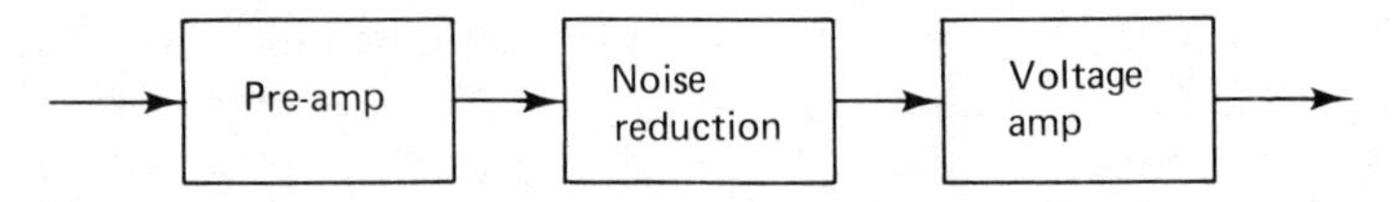

Figure 14-3 Block diagram for a tape deck with the addition of a Dolby noise reduction system. (Courtesy Zenith Radio Corporation.)

MECHANICAL SYSTEMS

Mechanical systems, in general, are much too complicated for newcomers to electronic servicing. The best bet for persons that are unfamiliar with these mechanisms is to take them to an authorized factory repair station for service. If these statements do not frighten you away from servicing these complex systems, obtain a service manual from the manufacturer of the system and very carefully follow the procedures detailed in the manual. The material presented in this section is intended to provide basic servicing information only.

Cassette system. Figure 14-4 shows the construction and layout of the cassette used for tape systems. The tape reels are used for both supply and take-up. The cassette can be operated in any physical position. The tape moves from the supply reel, past the guides, and on to the take-up reel. When this reel is full, the cassette is turned over and the tape moves from the full reel to the empty reel. Some systems require that this process be done manually by removing and inverting the cassette and then reinstalling it in the transport. Other systems have an automatic reversal unit built into the transport.

The openings in the cassette allow the tape heads to rest against the tape. These are provisions for both record/playback and the erase head. On the back side of the cassette there are two tabs. These are placed there for erase protection. A cassette that has a prerecorded tape will have these tabs broken off. When the tabs are missing, a mechanical lockout system in the transport keeps the record button from being used. Recording can occur only when these tabs are in place. In an emergency a piece of tape can be

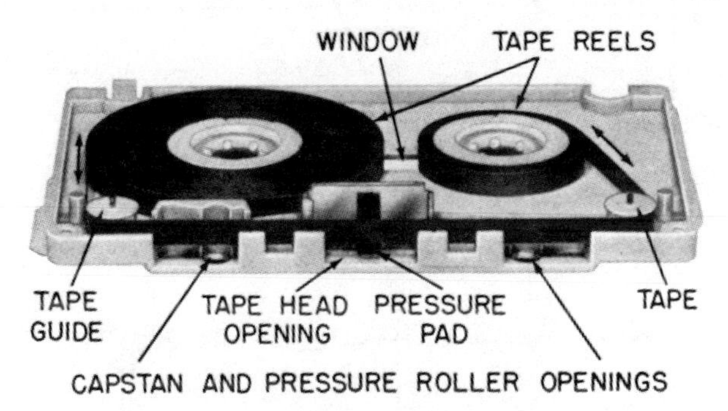

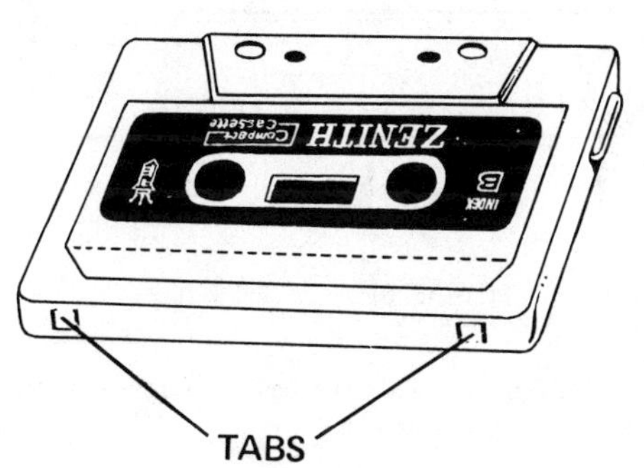

Figure 14-4 Construction of the cassette used for stereo systems. (Courtesy Zenith Radio Corporation.)

used to cover the holes for the broken tabs, thus allowing recording on the tape.

A view of the tape path past the heads is shown in Figure 14-5. The pressure roller is a part of the transport, as is the *capstan*. The capstan is used to provide a constant pressure on the tape. Speed is regulated in this manner. The capstan sits below the level of the tape in the transport. When the mechanism is engaged, the capstan is moved into the cassette and applies pressure to the tape and the pressure roller.

During playback or recording there is a buildup of tape oxides on the heads. This buildup affects the operation of the player. Often, it will

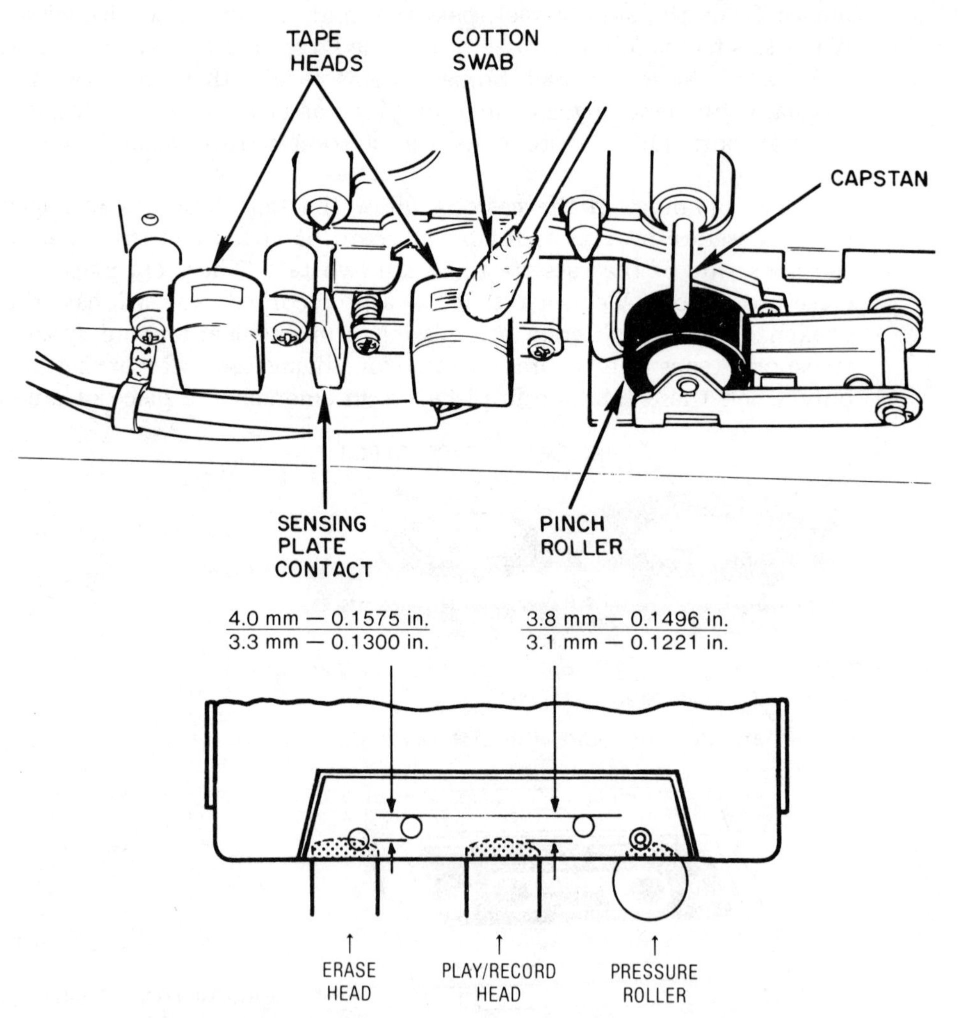

Figure 14-5 View of the tape path as it goes past the heads of the deck. (Courtesy Zenith Radio Corporation.)

cause distortion in the audio system or even mechanical problems to the transport. This figure shows one method of removing this oxide buildup. The process is designed to remove the loose oxide particles from the surfaces of the heads, the capstan, and the pressure roller. Follow the manufacturer's recommendation as to the type of chemicals to use for the procedure.

Be very careful not to scratch the surface of the heads with any sharp object. The resulting scratches will cause damage to tapes and shorten the life of the tape by causing unnecessary wear on it. This is also true for the capstan and the pressure roller. Often, an electric demagnetizer is used to neutralize any buildup of magnetic fields in the heads. Be careful not to bring the demagnetizer near prerecorded tapes or they will no longer have a valid recording on them.

A picture of the tape transport for a cassette system is shown in Figure 14-6. The two wheels used to drive the tape reels are shown. The cassette fits onto these. They are fluted to provide a solid connection for moving the

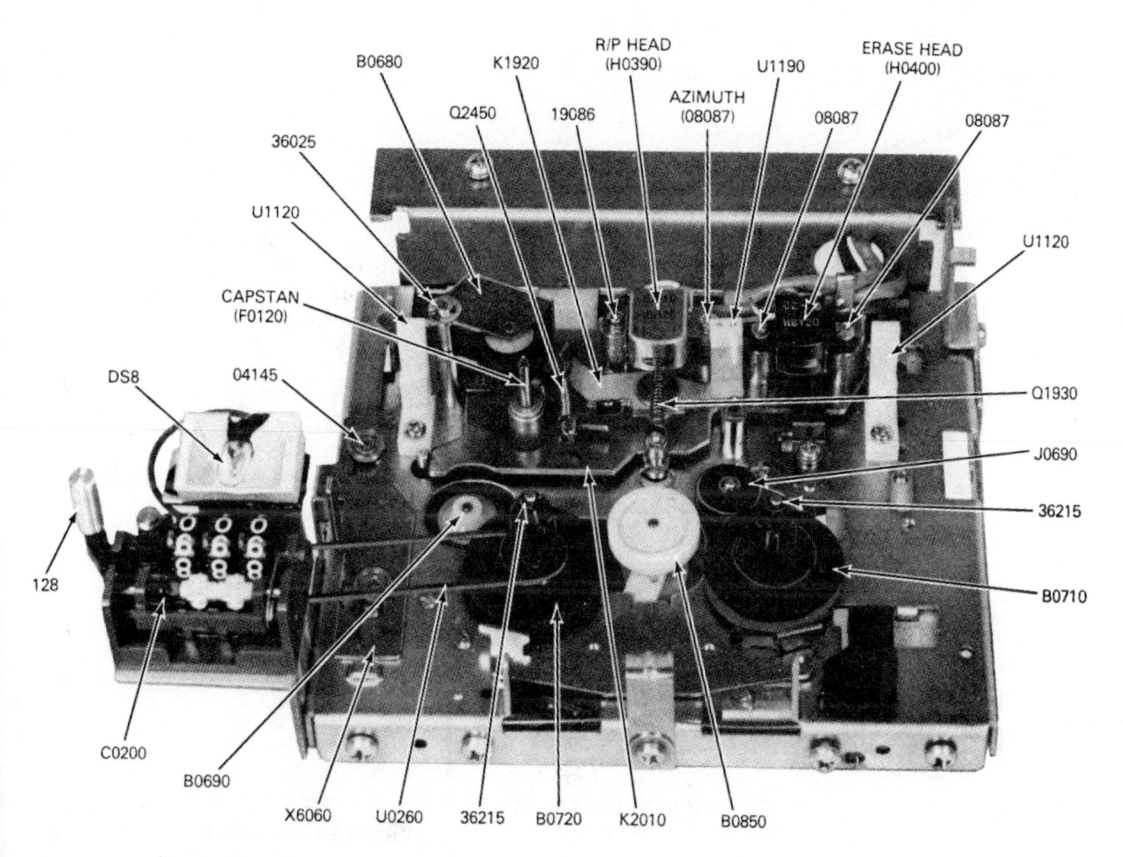

Figure 14-6 Mechanical features of a cassette transport system. (Courtesy Zenith Radio Corporation.)

tape. In addition to these, the record/playback head (R/P head) and erase heads are identified.

Cartridge system. The eight-track cartridge system uses a different method for moving the tape. This system is shown in Figure 14-7. A pressure roller is built into each cartridge. This roller applies pressure in order to hold the tape against the capstan shaft. The rotation of the capstan against the tape pulls the tape past the heads. This system uses an endless tape. The

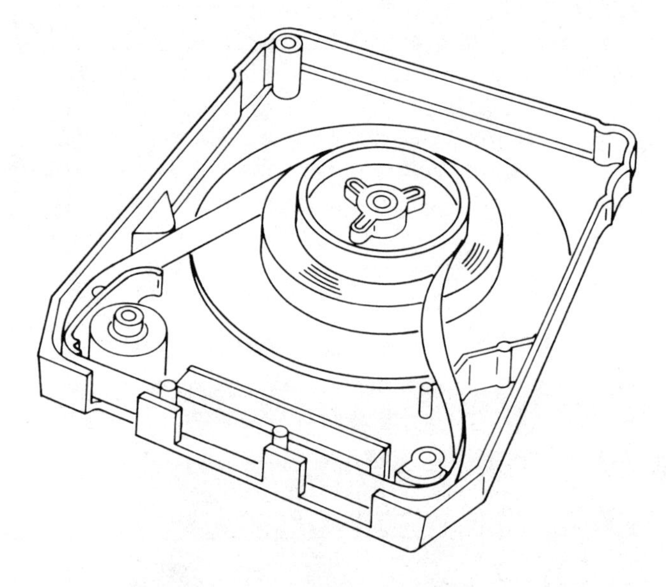

Figure 14-7 Construction of the eight-track cartridge used for stereo tape systems. (Courtesy Zenith Radio Corporation.)

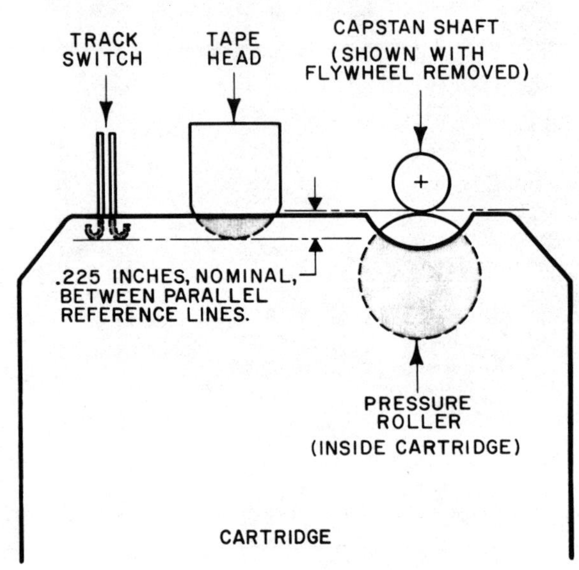

Figure 14-8 Layout of the mechanical portion of an eight-track tape transport. (Courtesy Zenith Radio Corporation.)

tape is constantly being rewound as it moves past the head and back into the cartridge take-up and supply reel. The mechanism for an eight-track transport is much simpler than that used for cassettes. The major moving part other than the capstan is the head indexing system. This device uses a stepping cam to move the head to its four different positions. Solenoids are used to pull the stepping mechanism and move the head. This system is shown in Figure 14-8. Head movement is accomplished in one of two ways.

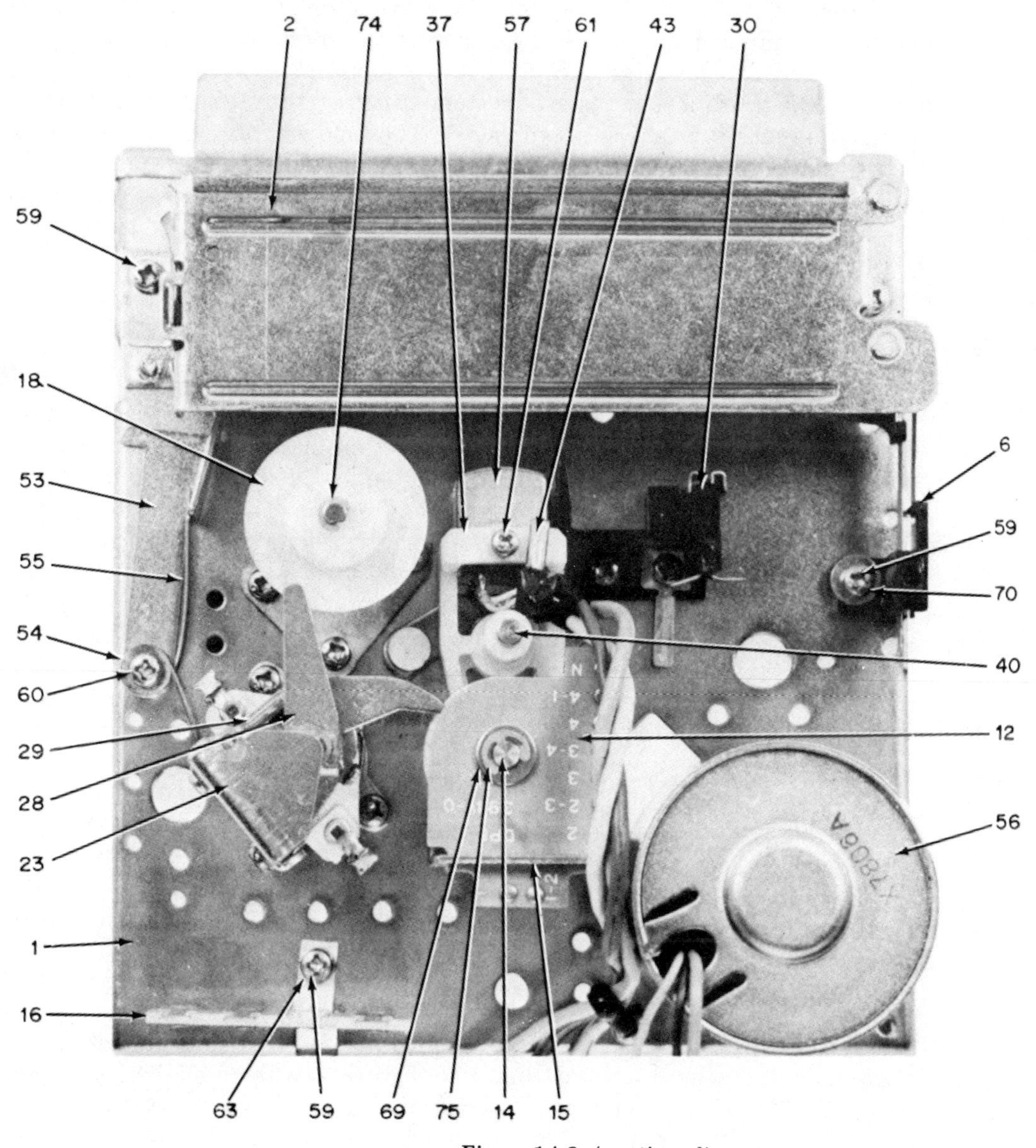

Figure 14-8 (continued)

Usually, there is a "track" switch located on the front panel of the unit. This switch activates the track change solenoid and rotates the track cam by 90°. A second method uses a set of contacts. These are located in the tape path. The tape ends are joined by a metalized strip. When this strip passes the track switch, it completes the solenoid engaging circuit and rotates the cam.

Reel-to-reel system. This system uses two separate reels for tape movement. It is much larger than its cassette counterpart. The system is similar to that used for cassettes except that the reels are not housed in a plastic holder as they are in the cassette system. Most of the reels used for tape recording are either 5 or 7 in. in diameter. The tape is threaded from one reel, past the heads and any tape guides, to the take-up reel. The process and problems found in reel-to-reel systems are very similar to those found in cassette systems. The exception to this is that the transport system for reel-to-reel systems is much simpler than its cassette counterpart.

ELECTRONIC SYSTEMS

The electronic section of the tape deck may be divided into five sections. These are the two amplifier channels, the bias oscillator, the noise processor, and the power supply. In addition to this a block called a *meter circuit* could be included in the system. A block diagram for one channel of a stereo tape deck is shown in Figure 14-9. This system shows separate heads for recording and playback. In many decks these are combined into one head assembly. The solid line areas on the diagram show those sections and signal flow paths used during the playback mode. Those areas that are shown by dashed lines are used during the recording process. This same system is shown in a combined schematic and block diagram in Figure 14-10. This figure is used to explain signal flow in the system.

Playback. The signal on the tape is induced into the tape head, where it is developed into a signal voltage. The signal is then fed to transistor amplifiers Q_{101} and Q_{103}. A multisection switch is used to switch between record and playback modes. This switch is shown in the playback position on the diagram. The signal is then fed from Q_{103} to the input to an IC. The IC contains audio preamplifiers and the Dolby noise reduction circuitry. The boxes marked "EQ" are frequency equalizer circuits. They are used to compensate for nonlinear frequency response in the recording tape.

Pin 5 of the IC is the input to a signal preamplifier. The output of this block is connected to a filter that is designed to attenuate all frequencies above 15 kHz. The signal goes from this filter to a second preamplifier. The output of the preamplifier is connected to the Dolby noise processing

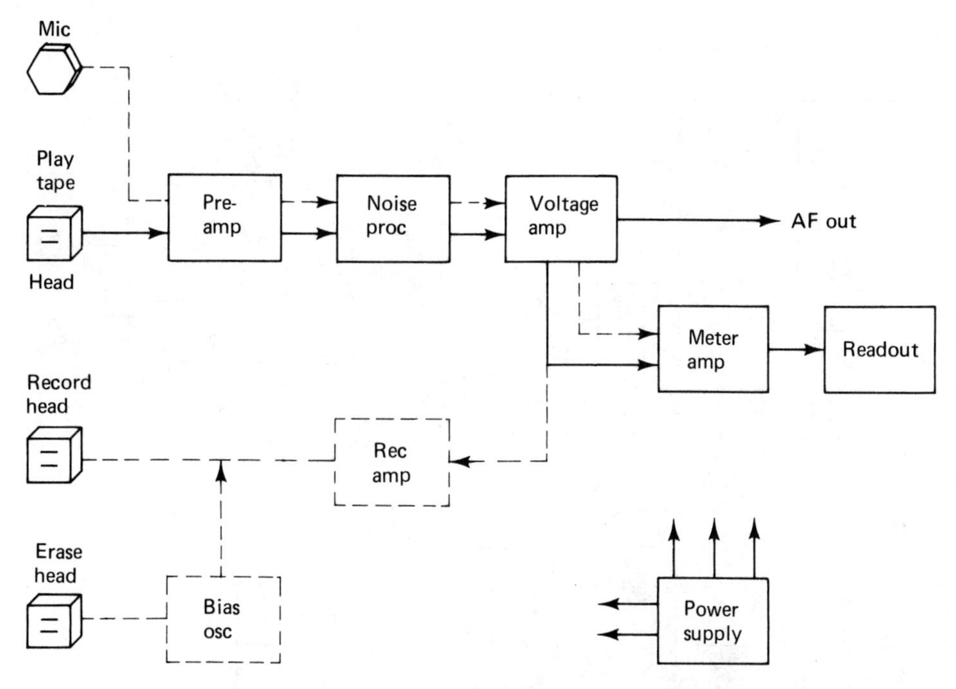

Figure 14-9 Block diagram for one channel of a stereo tape record/playback system. Solid lines show playback signal path and dashed lines show record signal path.

block. The output from the system is connected to pin 7 on the playback mode. It then goes to a buffer, or isolation, amplifier Q_{107} and then to the output jack of the deck. The output signal is also fed to a meter amplifier circuit.

Recording. During the recording process the input circuit to transistor Q_{101} is connected to a microphone or line in jack. The amplification that occurs on playback also occurs on record. In fact, the signal processing that occurs during "record" follows the same path as it does during "playback". The process and flow path through the Dolby portion of the IC also follows the same general path.

The output of the IC is switched to the record amplifier Q_{105}. Three different frequency equalizers are available for selection with the various types of oxides used in the manufacture of the tape. The signal then passes through a bias trap. This trap is used to prevent the signal in the bias oscillator from getting back into the record amplifier or other amplifier circuits. The bias oscillator serves two major purposes during recording. One of these is to provide a bias to the erase head. The second is used to apply an ac bias to the record head. A power supply voltage selection switch is used to control the amplitude of the bias signal.

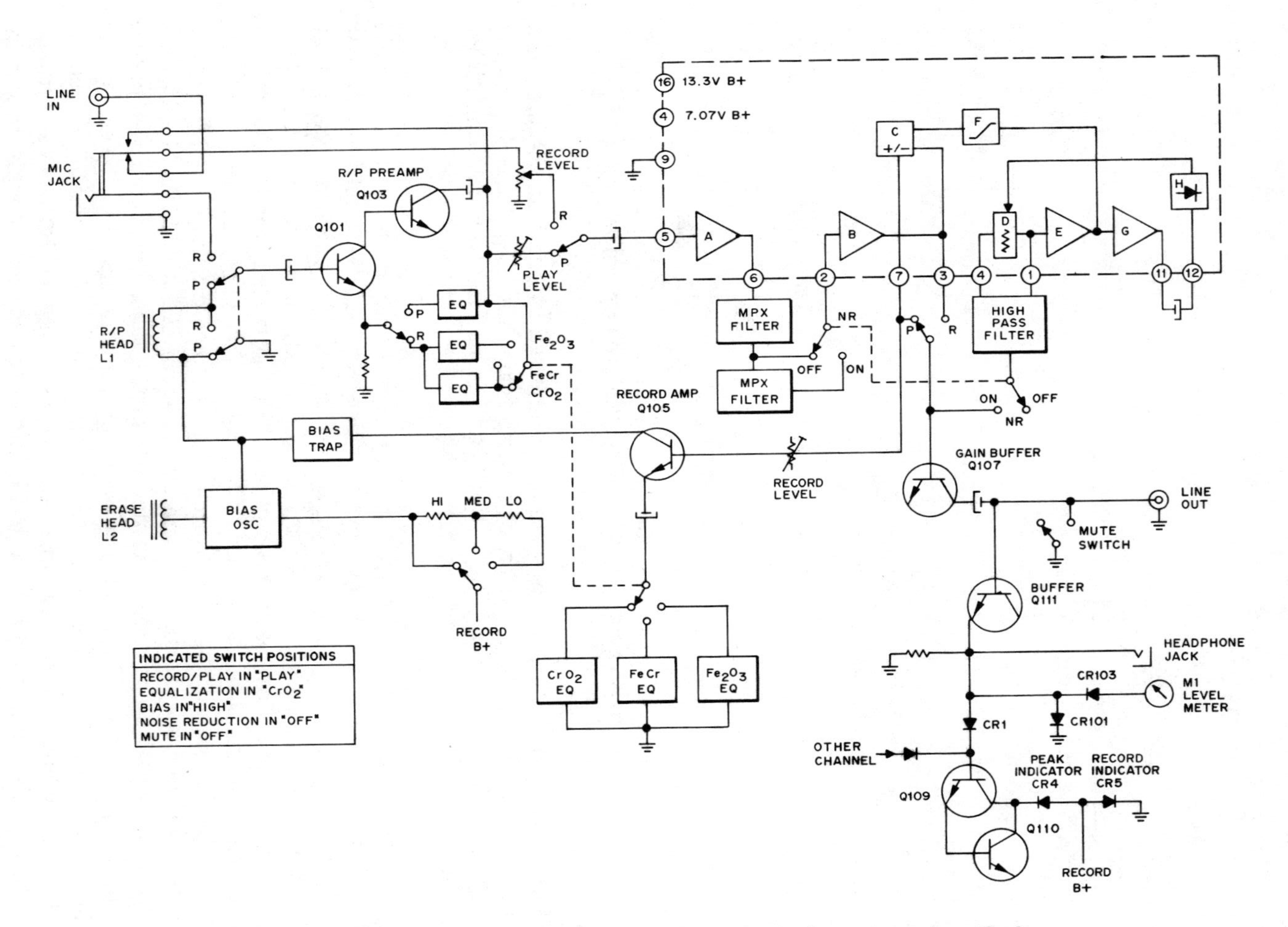

Figure 14-10 Combination block diagram and schematic for a stereo tape deck. (Courtesy Zenith Radio Corporation.)

Meter and indicator circuits. This system uses meters to indicate recording and playback signal levels. The ac signal is rectified by diodes CR_{101} and CR_{103} and then used for defection of the level meter M_1. In addition to this circuit there are two LED indicators. One of these, CR_4, is used to indicate excessive peaks of the signal. The second LED, CR_5, is used to indicate that the deck is in the record mode.

Power supply. The power supply section of a typical tape deck is shown in Figure 14-11. It has two secondary voltage sources. One uses a full-wave bridge supply to provide operating voltage to the motor and recording-level-meter amplifiers. The second circuit uses a full-wave center-tapped system and a series voltage regulator. This provides proper operating voltages to the electronics in the system. The biggest failure in either of these systems occurs with the electrolytic filter capacitors. When these fail, an ac 120-Hz hum will be heard in the electronic section. A failure in the meter or motor section will develop as poor motor speed or oscillating in the meters.

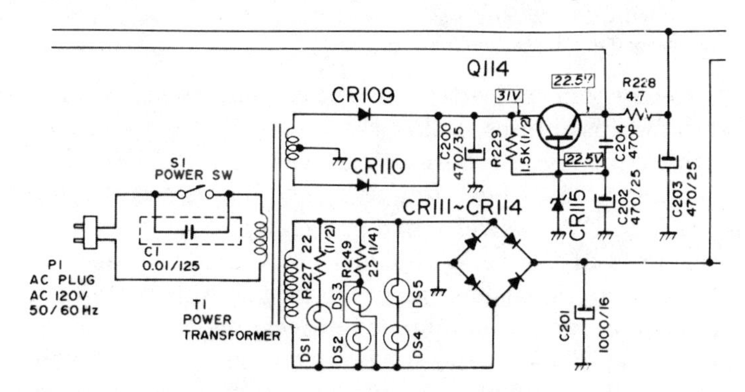

Figure 14-11 Schematic diagram for a stereo deck power supply. (Courtesy Zenith Radio Corporation.)

Bias oscillator. One bias oscillator is used for both channels in this system. A schematic for the bias oscillator is shown in Figure 14-12. This system uses a single transistor oscillator, Q_{113}. It oscillates at a frequency of 105 kHz. This frequency may be different in other systems. A switch, S_{104}, is used to change the amplitude of the bias signal. It will range from a low of 25 mV to a high of 40 mV, depending on its position. The variations are done by changing the level of operating voltage in the oscillator system.

Noise reduction system. The most popular method of reducing noise in a tape deck seems to be the Dolby Laboratories system. This system is both an encoder and a decoder. It functions by boosting low-level, high-frequency program material during the recording process. When playing this material

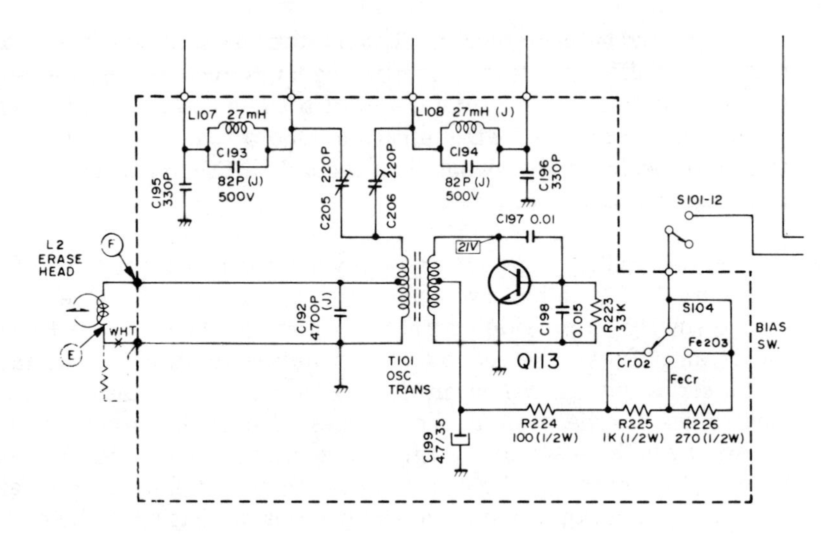

Figure 14-12 Bias oscillator section of a tape recorder system. (Courtesy Zenith Radio Corporation.)

back, the system will attenuate these frequencies. This method will reduce background hiss noises during the record and playback modes. These noises are high frequency in nature and are virtually eliminated with this procedure. The same system of encoding and decoding is used to improve the signal-to-noise ratio in FM receiver signal processing.

A graph of the type of signal processing that occurs is shown in Figure 14-13. A loud signal has no processing in this system. A medium-strength signal has a moderate amount of processing and a weak signal receives a lot of processing. This is true for both recording and playback. The chart also shows that the frequency at which this occurs varies with the strength of the signal.

A block diagram of the Dolby system is shown in Figure 14-14. The signals that require encoding are processed in the left-hand box. These are added to the original signal to enhance the high-frequency portions. The

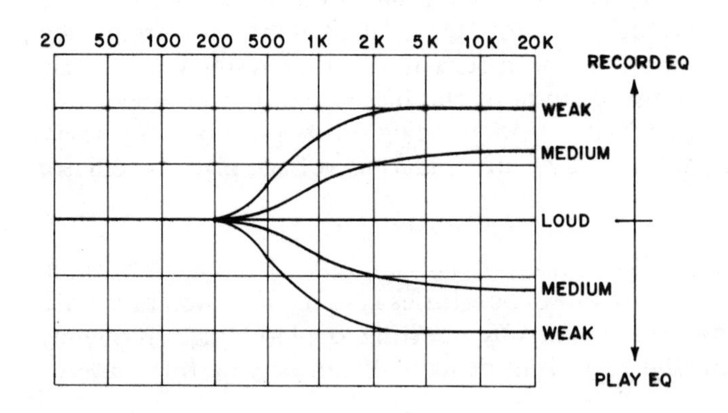

Figure 14-13 Graph of the signal levels processed by a noise reduction system.

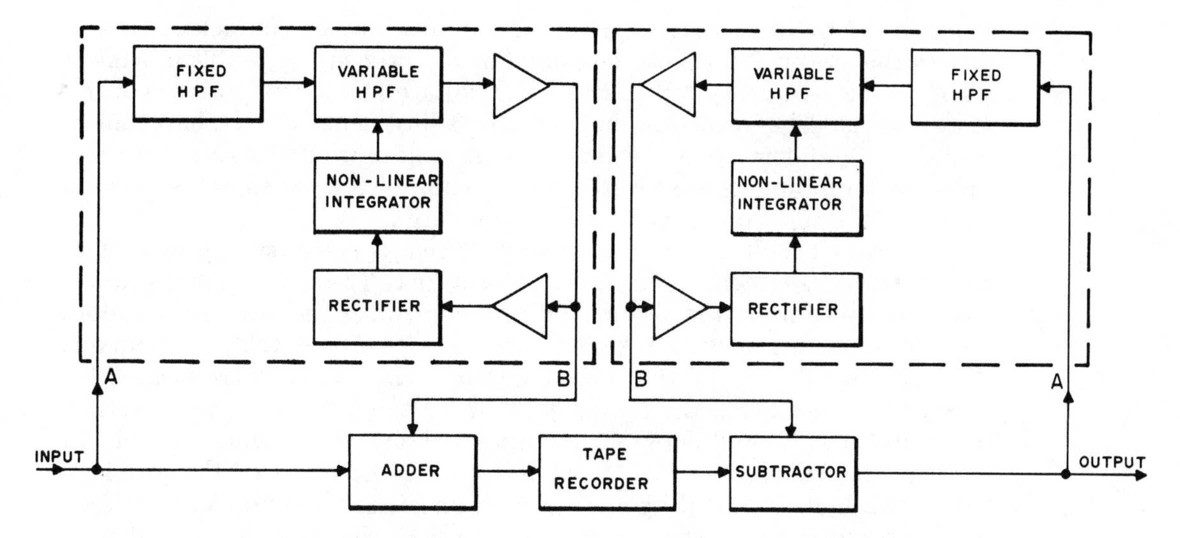

Figure 14-14 Block diagram for a Dolby noise reduction system. (Courtesy Zenith Radio Corporation.)

output of the system is fed to the tape recorder and becomes a part of the recording on the tape. During playback the signal from the tape is processed by subtracting the high frequencies from the recorded signal. This restores the information to its original format without the annoying high-frequency hiss usually associated with tape playback.

TROUBLESHOOTING TAPE DECKS

The first step related to repair of the tape deck is to attempt to isolate the problem. This, as is true for any electronic device, can be accomplished by trying to operate the machine. Use a good test tape and try to play it. The analysis should include checking each channel for equal output. Of course, one must have a test tape that has equal signals recorded on both channels. If the machine plays back the tape successfully, the problem is elsewhere. The next thing to do is to try to record on the tape. Use a tape that has been previously recorded. If the new recording is clear, the bias oscillator is working properly. When this oscillator fails, the new material is recorded over the old material and both are heard at the same time. Observe the recording-level meters during the recording process. Both meters should function equally well. If one does not indicate a level that is equal to the other, the problem may be in the meter amplifier circuits and not in the record or playback part of the system.

While the playback or recording tests are being conducted, one should look and listen for other problems. One problem that often occurs is that the speed of the tape is incorrect. This may be caused by dirt or lack of lubrication in the motor. It may also be caused by a power supply problem. Almost all of the better-quality tape decks use a voltage regulator circuit to control motor speed. The regulator may require adjustment or may have failed. Additional problems may be due to mechanical troubles. Are there any squeaky or rubbing sounds heard in the machine? These often can be located by a careful inspection as the machine is operating.

Another problem in any electronic system is power supply hum. This will be heard as either a 60-Hz or a 120-Hz tone. The problem often can be localized by rotating the volume or level control. If the level of hum does not change as this control is rotated, the hum is being developed in circuits that are not affected by the volume control. These circuits are located between the control and the output jack. When the hum level can be varied, the problem is located between the input jack and the volume control. An oscilloscope will be a valuable aid in locating the cause of the problem. Often, this hum is caused by an improper or missing common lead between the deck and other electronic devices used with the deck. A fast test is to use a jumper wire and connect it between the chassis of both units. If the hum level drops or disappears, look for an open common, or ground, lead.

Another common complaint is a high-frequency noise or hum. This can be caused by a poor common connection between the circuit board and the chassis or case of the deck. Often, one of the level meters will oscillate. This is a good clue as to which channel has the problem. Transistors can also produce noises. The type of noise heard can be a popping or frying sound. Transistors can be tested by removing them from the circuit, applying heat or cold to its body, or by shorting base and emitter leads together. The least desirable approach is to remove the transistor. This takes up expensive service time and may produce other problems. Shorting base and emitter leads puts the transistor into cutoff. It acts like a very high resistance under this condition. If the noise disappears, the trouble is located. A Freon spray can be used to cool a suspected component. When the noise occurs only after the set is on for a while, heat may be a problem. A specific component may have changed value due to the normal rise in temperatures in the unit. Spraying this chilling gas on individual components will help to locate the problem part.

There are times when a motor becomes noisy. The noise may be either mechanical or electrical. Mechanical vibrations often produce noises in transistors. Manually stopping the motor will help to determine if this is the problem. Motors also have a tendecy to arc. Connect a 500-μF capacitor across the motor leads. If the noise stops, add this capacitor to the circuit. If the noise does not stop, the motor will have to be replaced.

Another area of trouble in a tape deck is in the mechanical switching circuits. Switch contacts often become dirty or oxidized. When contacts develop these conditions they have also added a higher than normal resistance to the switching circuit. The extreme result is a total loss of ability to operate the system. In many cases a partial loss of operational control occurs and the system may even be intermittent. There are several excellent contact cleaners on the market today. Proper application of one of these will clean the contacts and restore the system to normal operation.

When the problem is related to the electronic circuits, the normal rules for troubleshooting apply. Most of the amplifier systems use a linear signal flow path. An oscilloscope is used to locate the specific area of trouble. Once the problem area is identified, both the oscilloscope and an electronic meter are used to locate a specific component that has failed. The procedures for troubleshooting and repair of a tape deck electronic system are very similar to those used for any other amplifier.

As a part of the test and adjust procedure, the frequency of the bias oscillator is measured. This is measured with either a frequency counter or an oscilloscope. The digital readout of the frequency counter makes this tester easier to use than the oscilloscope for this purpose. Adjustments for the frequency and amplitude of the bias oscillator signal are made according to the specifications established by the equipment manufacturer.

QUESTIONS

14-1. Name three types of tape systems presently used for stereo reproduction.

14-2. What block is used only for recording?

14-3. What signals are developed in the extra block, and how are they used?

14-4. What is a Dolby system and how is it used?

14-5. Why is it necessary to clean and demagnetize tape heads?

14-6. Why is a recording-level meter used?

14-7. Which component in the power supply has a tendency for early failure?

14-8. Why is it necessary to change the amplitude of a bias/erase signal?

14-9. Which block fails when double recording occurs?

14-10. Which sections and components have failed when one hears a low-frequency hum in the system?

Chapter 15

Some Topics of General Interest

There are many facets of troubleshooting information. They are often so limited in technical information that they do not warrant a separate chapter in this book, but they are too important to omit. These topics are presented in this final chapter. They are included to make sure that the technician just starting into the repair of stereo systems has a broad background in the major factors involved in stereo equipment servicing.

SAFETY PRECAUTIONS

The service technician has to realize that his or her role is to *repair* and not to modify equipment. The manufacturer has spent great sums of money in the development of a sophisticated piece of equipment. Engineers are employed to design this equipment. Unless there is a factory bulletin showing a modification, the equipment is to be repaired to restore the integrity of the existing circuitry. The technician who modifies a circuit opens the door to lawsuits and makes himself liable if the modifications have caused damage to the unit or anything or anybody using it.

All service work should be performed only after the technician is familiar with all the recommended safety checks and guidelines established by the manufacturer. One of the first tests relates to fire and shock hazards. This is a "looking" test. Carefully examine the unit, looking for possible short-circuit conditions. All components should be in a position where they cannot cause a short circuit. This examination should consider what will occur when the cover, case, and bottom chassis pan are installed. The possibility of moving a component is also very likely if the unit has been carried to the service facility.

When removing guards or protective covers from a piece of equipment, note the location. Be sure to reinstall these guards. Also be sure to reinstall any insulators or insulating materials that were removed during the repair process.

Part of the looking process should include an inspection of the ac line cords. Look for open or cracked insulation or any fraying of wires.

Be extremely careful to replace fuses and any other special parts with exact replacements. Many manufacturers indentify special components on the schematic diagram. Some examples of this are shown in Figure 15-1. These components often are flameproof resistors. The flameproof resistor is one that will not burst into flame when it fails. Other components that may have a low safety factor unless replaced with exact factory parts also fall into this category.

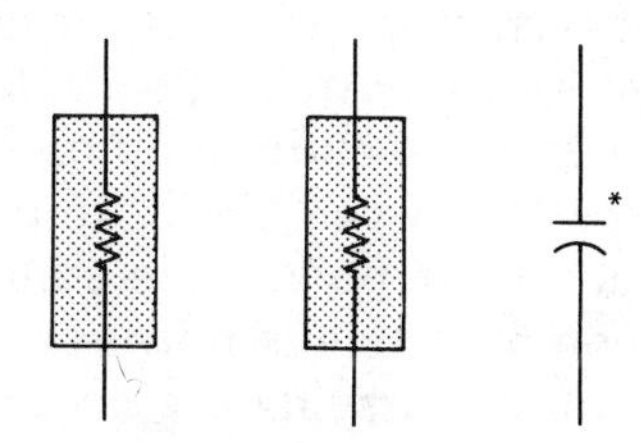

Figure 15-1 Many schematic diagrams identify critical components by one of these methods.

Another very important test is the ac leakage test. This test is performed on exposed chassis parts. These parts include knobs, antenna terminals, and ground, or common, terminals. The purpose of this test is to make certain that there is no ac current path from the unit to the normal ac earth ground. The presence of an ac path between these points is a high shock hazard.

The procedure for setting up this test is shown in Figure 15-2. The unit to be tested is connected by its ac line cord directly into the ac power line. An isolation transformer is not used for this test. In fact, the use of an isola-

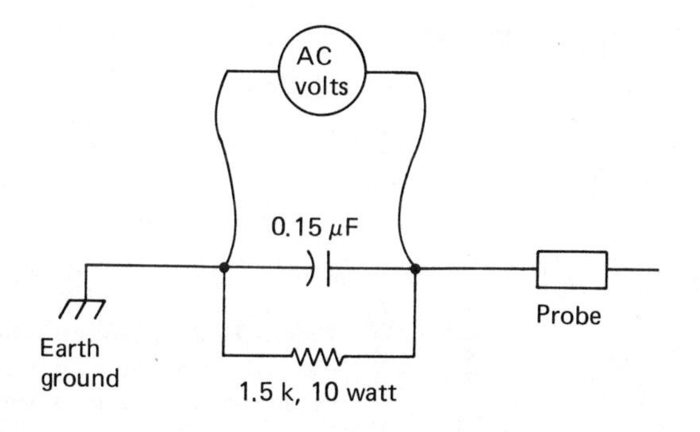

Figure 15-2 Safety checks for ac leakage require this circuit. It can be built by any technician.

tion transformer will void the results of the test. A voltage is measured across the parallel wired 15,000-Ω 10-W resistor and a 0.15-μF ac capacitor. These two components are connected between an earth ground and become a part of the probe used for the test. The probe is moved to each exposed part of the set and the ac voltage read on the meter. The ac power line plug is reversed and the entire procedure is done again. If any one voltage reading exceeds 0.75 V rms, the set is a potential shock hazard. The 0.75 V rms is equal to 0.5 mA of ac current ($I = E/R = 0.75/1500 = 0.5$ mA). This is the maximum ac leakage allowable under safe operating conditions.

GROUND LOOPS

A ground loop is an undesirable circuit common path. This situation occurs when an audio cable is used to connect 60-Hz power circuits between two circuits. It also occurs when one attempts to use the audio cable to carry 60-Hz power current between two independent chassis or units. This situation is illustrated in Figure 15-3. An audio cable is shown connecting the output of the signal source to the audio amplifier. The cable is designed to carry low-level audio signals between these units. Each of the units is designed to have its own ac 60-Hz power line connection to ac common. If, for any reason, this ac common line is not used, or not installed properly, the audio cable has the job of carrying the ac common current. The audio cable is not designed to do this type of work.

The major problem when attempting to use an audio cable as an ac common line is that the result is an introduction of the ac power line signal into the audio circuits. Since the ac power line frequency is in the audio-frequency spectrum, the result is a 60-Hz hum. This hum is amplified and carried throughout the system. It produces an unwanted noise in the system. The best method of reducing or eliminating this problem is to be certain that there is a better path for the ac common than the audio cables. Power-line cords should be polarized or connected so that the same side of the ac line is common to all units. In addition, a separate ac common or ground

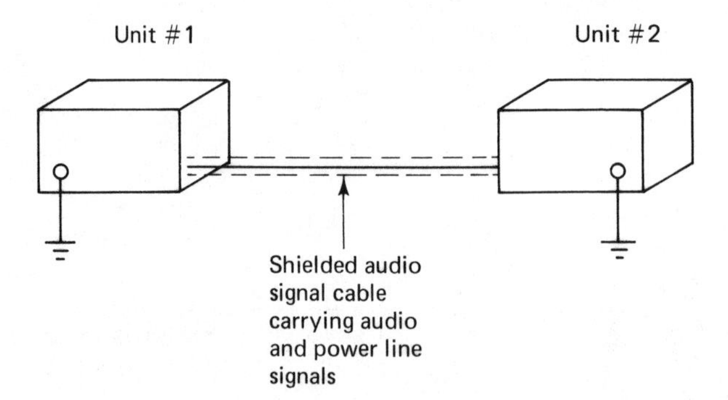

Figure 15-3 A ground loop, or unwanted signal-carrying path, will introduce ac hum into the system.

wire should be connected between all units. If these procedures do not eliminate the problem, and it is not a result of ac leakage, the problem is in the power supply of one of the units and is not a ground-loop problem.

INTERFERENCE PROBLEMS

One repair problem that rarely occurs on customer equipment when it is in the repair shop is interference. This interference is caused by radio transmitters, electric motors, industrial electric equipment, electrical appliances, lamp dimmers, or any one of several other noise-generating types of equipment. The term used to describe these types of interference is *radio-frequency interference* (RFI). The solutions to the problems caused by RFI are about as great in number as the causes of the interference. One of the most comprehensive manuals on this subject is available from the Electronic Industries Association.* It is entitled *Interference Handbook, Audio Rectification.* It, and a companion manual on TV interference, should be read by all service technicians.

The major problem with RFI is that it can enter the system at any point between the antenna and the speaker. The first thing the service technician should attempt to do is to localize the problem. This will include determining when the problem occurs. This will provide a clue as to whether a certain piece of equipment is radiating RFI. The time of occurrence also helps to determine if the RFI is produced by a daytime-operated business or by a hobbyist in the evenings or on weekends. It will also help identify if the problem is localized as a specific piece of electronic equipment. Often, the cause of the problems in the home can be localized by unplugging suspected equipment. If this stops the problem, the cause has been located.

One reason for the problem of RFI is that manufacturers of electronic equipment cannot possibly provide the adequate shielding or filtering required to solve each and every RFI problem. The cost of this additional circuitry would problably raise the unit cost so high that most of us would not be willing to buy it. Manufacturers will, therefore, make a reasonable attempt to install shielding or filtering that solves the majority of problems. The rest is up to the service technician on an individual basis.

A solution that seems to solve most RFI problems is to add an ac line filter to the system. This usually consists of the type of filter shown in Figure 15-4. A pair of capacitors are connected between each side of the ac line and circuit common. The size of these capacitors will vary depending on the RFI cause. A good value to use in many situations is an 0.01-μF capacitor. Be sure that the voltage rating of the capacitors is high enough to avoid breakdown. A recommended value for 120-V circuits is 400 V. An

*Consumer Electronics Group/Electronic Industries Association, 2001 Eye Street N.W., Washington, DC 20006.

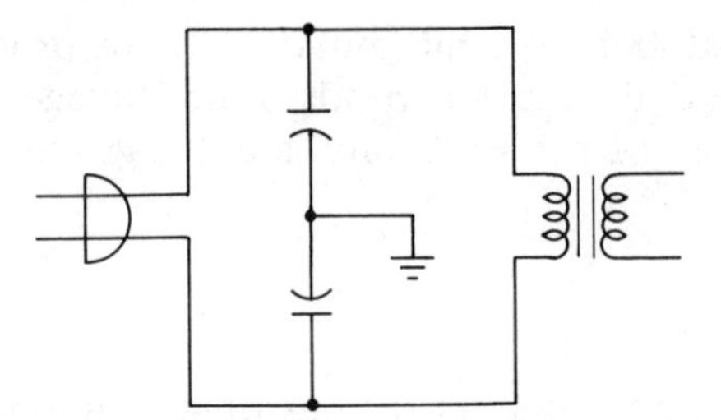

Figure 15-4 Ac line filter systems use small capacitors to bypass noise spikes.

alternative solution to power line RFI problems is to install one of the many excellent commercial filter units that are available today.

EQUALIZER CIRCUITS

The ideal reproduction level of sound produces an equal level of response over the entire audio-frequency spectrum. This ideal response is not practical because the recording process does not use a flat response curve when records or tapes are produced. The recording process requires a larger signal level for high frequencies in order to reduce high-frequency background noise on records or tapes. The method of doing this is to boost the level of the high frequencies during recording. A block system called an *equalizer* is used to do this. In reality, the equalizer is a filter that enhances the higher audio frequencies. The lower frequencies are attenuated during the process. A response curve used for recording is shown in Figure 15-5. The upper drawing shows the recording response curve. The second, or lower, drawing shows the response curve used for playback. This curve is opposite to that used for recording. The playback system enhances the low frequencies and attenuates the high frequencies. The net result is a reasonable flat response for all audio frequencies. The system standard for these curves is called an

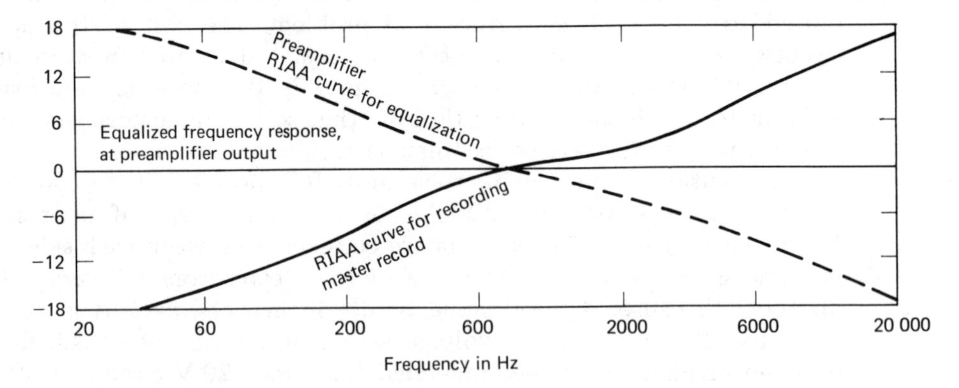

Figure 15-5 Response curve for phonograph records. The upper curve is for recording and the lower curve is for playback.

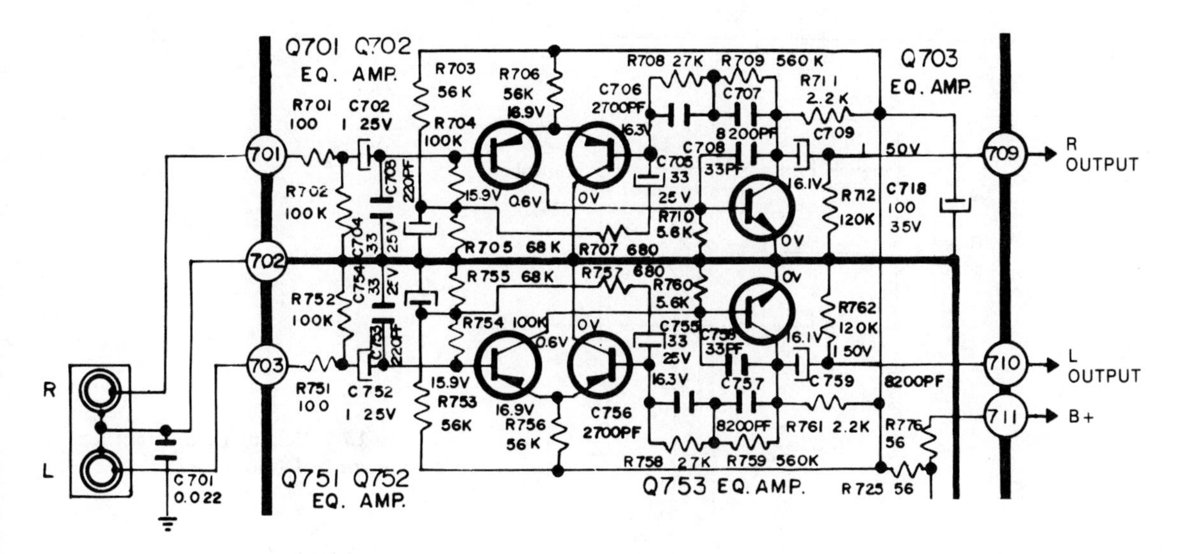

Figure 15-6 Schematic diagram for an equalizer network for phonograph records. (Courtesy Zenith Radio Corporation.)

RIAA curve. The system was developed by the Recording Industry Association of America (RIAA).

There are different circuits used for records and for tape players. The curves shown are for records. The circuit in Figure 15-6 is a typical RIAA record playback equalizing network. It consists of a group of resistors and capacitors in a series-parallel connection. Resistors R_{708} and R_{709} with capacitors C_{706} and C_{707} are the major components in this network. It is connected between transistors Q_{702} and Q_{703} for the right channel of this stereo preamplifier.

TONE CONTROLS

The purpose of the one control system is to permit the listener to adjust the frequency response of the system. A tone control system can be used to either enhance or attenuate both high and low frequencies. A typical tone control system is shown in Figure 15-7. This system uses variable resistors and fixed-value capacitors. This changes the frequency response of the system. There is a loss of signal as it is processed through this system. The signal loss is made up by additional amplification in the system.

Component failure for this system is most often due to a failure of the variable control. One problem is that the contact arm becomes dirty and develops increased resistance. The second problem is due to an open common lead in the system.

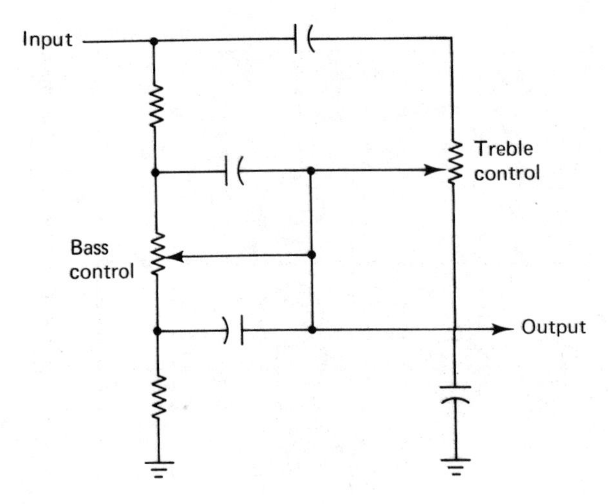

Figure 15-7 Schematic diagram for a tone control system.

SUMMARY

The emphasis on repair of almost any electronic problem is based on a thorough knowledge of how the system works. This knowledge, in addition to an understanding of electron current flow and basic circuits, will enable the technician to repair almost all systems. One must know how to use electronic test equipment. The knowledge of which piece of equipment to use and how to evaluate the results of the tests is essential if repairs are to be made quickly and correctly.

Techniques for successful repair are developed as one gains experience. The best method of approaching any service problem is to develop a standard approach to all service problems. Use a standardized format, or checklist, to evaluate the system. Valid tests include operation of the system and the resulting elimination of all properly working sections. The area that is now left is the area in which the problem is located. Once this is done, the use of test equipment will locate the specific component that is defective.

The emerging service technician must keep up on current circuits and systems. There are several ways to do this. These include attending school, attending service seminars, and reading about circuits in the many excellent books and magazines that are available. The final suggestion is to join a professional organization. There are local, state, and national professional organizations. These will provide a feeling of professionalism to the service technician in addition to providing information on current problems or topics of interest to the technician.

QUESTIONS

15-1. Why is it wrong to reengineer or modify circuits when repairing units?

15-2. What are visual signs of a bad line cord?

15-3. Why are flameproof resistors used?

15-4. What is an ac leakage test?

15-5. What is a ground loop?

15-6. What is RFI, and how does it affect the stereo system?

15-7. What is an equalizer, and how is it used?

15-8. What effect do tone controls have in the frequency response of the system?

15-9. Why is it necessary to use a record equalizer?

15-10. How does a short circuit between an operating circuit and a metal chassis affect operation of the stereo unit?

Index